In Search of Lost Time

In Search of Lost Time

Derek York

University of Toronto

 Routledge
Taylor & Francis Group

LONDON AND NEW YORK

First published 1997 by IOP Publishing Ltd

2 Park Square, Milton Park, Abingdon, Oxfordshire OX14 4RN
52 Vanderbilt Avenue, New York, NY 10017

Routledge is an imprint of the Taylor & Francis Group, an informa business

First issued in hardback 2019

British Library Cataloguing-in-Publication Data

A catalogue record for this book is available from the British Library.

Library of Congress Cataloging-in-Publication Data are available

Typeset in TEX using the IOP Bookmaker Macros

ISBN 13: 978-0-7503-0475-7 (pbk)
ISBN 13: 978-1-138-42977-2 (hbk)

To Lydia, Link, Katherine and Joseph

Contents

Preface

For over a third of a century my life has been spent in search of lost time. In 1957, Professor L R Wager, head of the Department of Geology at Oxford, invited me to join a small group of scientists who were engaged in setting up radioactive dating techniques in his laboratory. Having just graduated in physics, this seemed like a large step into the unknown. However, luckily, I accepted his invitation and have ever since been in pursuit of thousands and millions, and thousands of millions, of years. This book is a reflection of my obsession with time and its measurement. It will take you from the pyramids of Egypt, through Stonehenge, the North China plain, to the universities of Cambridge, McGill, Chicago and Toronto, to the Patent Office in Berne, and back to the Ethiopian desert on the banks of the Awash River. On this time-odyssey you will enter the mind-bending universe of the special and general theories of relativity, the ghostly world of quantum mechanics and the unpredictable haunts of chaos. You will have good companions to share and illuminate the path—from Jonathan Swift's *Gulliver's Travels* and Lewis Carroll's *Alice in Wonderland* to J B Priestley's *Dangerous Corner*. You will meet the father of master-spy Kim Philby in the Empty Quarter of Arabia, the fantasist Velikovsky in the clouds, and Newton, Darwin, Rutherford, Einstein and the great earth scientists of this century who fathomed the depths of lost time and discovered the age of the earth. May you enjoy your search for lost time as much as I have mine.

Derek York
Toronto, May 1997

Acknowledgments

I would like to thank my wife Lydia and son Link for their continuing encouragement during the writing of this book. Many of my ideas were shaped in discussions with them. Terry Christian of *The Globe and Mail* was a pleasure to work with and is a great friend. Ruth Bobbis and Maria Wong did an excellent job of typing the manuscript and were helpful in many ways. Sarah Peat was a great help in finding illustrations and obtaining permissions. Peter Binfield and Joanna Thorn of Institute of Physics Publishing sustained me with their enthusiasm for the book and their efforts during its final production.

Finally, I would like to thank Mike and Sorel Garvin for making their beautiful cottage on the beach at Clifton, Capetown available to us. Their generosity and the siting of the cottage on the Cape Granite (which crystallized at the dawning of the Cambrian period) inspired the planning and the writing of a significant portion of *In Search of Lost Time*.

I gratefully acknowledge permission to use illustrations from the following sources:

Figure 1.6 by courtesy of The George Crofts Collection, Royal Ontario Museum.

Figure 5.1 is a computer graphic by Khader Khan and Raoul Cunha.

Figures 7.2 and 7.3 by courtesy of the Institute of Human Origins, Berkeley.

Figure 8.1 by courtesy of Richard Grieve, Natural Resources Canada.

Figure 16.1 by courtesy of the Tate Gallery, London/Art Resource, New York.

Chapters 6, 8, 9, 10, 12, 13 and 14 were based in whole or in part on articles by the author which have appeared in *The Globe and Mail*, Canada. Chapter 5 is based on an article by the author which appeared in the American Geophysical Union's *Eos*. I am grateful for their permission to use this material.

CHAPTER 1

THE PYRAMIDS, STONEHENGE AND THE CHINESE ORACLE BONES —SAME TIME NEXT YEAR

Throughout his existence in what we might call 'civilized' conditions, man has been obsessed with two contrasting aspects of time: its cyclic repetitive nature on the one hand (as exemplified by the rising of the sun above the horizon every day, its setting each night, the monthly waxing and waning of the moon, the return of the seasons, the yearly flooding of the Nile, etc), and its linear non-cyclic non-repetitive nature on the other (as represented most dramatically by the one birth and one death that we all experience).

We see this same contrast in modern cosmologies, where some see infinitely cycling universes with endlessly repeating space and time (see chapter 16), while others envisage a universe born only once, expanding for ever into maximum disorder. In this chapter we look at our earliest struggles to come to grips with these ideas, and our efforts to mark the passage of time, through the testimony of the pyramids of Egypt, Stonehenge and the Chinese oracle bones.

The pyramids reflect both these sides of time's coinage. The Egyptian kings sought to defeat the grasp of death and hoped that the pyramids (figure 1.1) would protect their bodies against the linear ravages of time and enable them to return to some form of life, perhaps like the sun on the first day of spring, the vernal equinox. Remarkably, they oriented their pyramids almost perfectly to face the first and last rays of the sun on this day of renewal.

Figure 1.1: *The pyramids of Giza.*

In contrast, we shall see that, rather than marking the equinoxes, the builders of Stonehenge marked midsummer's morning and the metronomic motions of the sun and moon along the eastern and western horizons during the year.

Finally, where the Egyptians and ancient Britons built massive stone monuments to time the movements of the sun and moon, the Chinese timed the earth's rotation with a simple stick and obtained results in extraordinary agreement with modern values.

The pyramids of Giza

The pyramids of Egypt, about 80 of which are known, lie scattered along the fringe of the desert to the west of the River Nile. They were the product of the so-called Old Kingdom which stretched from roughly 2700 to 2200 BC. The pyramids were tombs of kings and queens and resulted from the great concern of the Egyptians with the idea of life after death.

During the Pyramid Age there were two dominating religious cults: the sun cult with its sun god Rē-Atum, and the cult of the god Osiris. The sun cult was derived from a powerful group of priests at the city of On which was later called Heliopolis (City of the Sun) by the Greeks. The Egyptians were inconsistent in their beliefs. Several explanations, for instance, existed for the sun's daily motion across the sky. The most delightful idea involved the scarab beetle (dung beetle). I E S Edwards has pointed out that the Egyptians had often seen the dung beetle rolling a ball of dung along the ground before depositing it in a hole. They believed that a young scarab beetle then eventually emerged from this ball by some process of self-generation. Since they viewed the sun as the source of all life, they associated the sun with the dung ball and the sun god, who pushed the sun, with the scarab beetle who propelled the dung ball. Edwards notes wryly that the dung ball watched by the Egyptians is merely a reserve supply of food, while a pear-shaped ball of dung containing an egg is kept hidden by the mother until hatching occurs. The sun god was called Khepri in this scarab form.

Rē-Atum was principally the god of the living. Osiris, in contrast, was considered as the god of the blessed dead and the region of the dead. However, these deities shared one property of fundamental importance to the Egyptians: that of survival after death. Thus in the legend of Osiris he is murdered by Seth and yet rises again through Isis's magical intervention, while Rē-Atum's daily disappearance at sunset was regarded as his death, which was followed every morning by his rebirth in the east. Following these divine examples, the Egyptian kings hoped to enjoy a similarly pleasant life after death. To this end they thought it important to preserve the dead body and, together with it, they entombed valuable possessions and food.

About 8 km west of Giza, not far from Cairo, stand three enormous pyramids: the tombs of the Kings Cheops, Chefren and Mycerinus. In the words of Edwards, 'Together, these three pyramids constitute possibly the most celebrated group of monuments in the world.' The pyramid of Cheops is the oldest and the most impressive. Its square base occupies over 4000 m^2 while its apex originally soared 147 m above. Later Egyptians removed the limestone cover from the faces and other blocks are missing from the top, so that now the topmost point is about 137 m above the base. The deep core of the pyramid is a massive chunk of rock. However, the main body of the pyramid is built of blocks of local stone. The original surface dressing was of limestone from Tura. It has been estimated that 2300 000 blocks were used, of average mass 2500 kg. Some blocks reached an estimated 15 000 kg. The ancient Greek historian Herodotus stated that he had been told that the stupendous task of assembling the Great Pyramid (the pyramid of Cheops) occupied 400 000 men annually for 20 years. The British archaeologist Petrie estimated that 100 000 men would have sufficed. Regardless of the details, however, the result was truly remarkable.

Access to the pyramid was gained through an entrance in the north face, as was traditional. After the burial ceremony, this entrance was plugged with massive granite blocks and then most probably hidden by carefully placed limestone slabs. Such precautions were routinely taken to protect the body and possessions of the king from looters. This was to no avail, however, as Cheops' body was not found by modern investigators and his pyramid had been plundered, no doubt being violated several times.

This greatest of all pyramids is of fundamental interest to us because of the following fact of observation: Cheops' pyramid is oriented with extreme accuracy to the north–south and east–west directions. The detailed measurements by J H Cole of the Survey Department of the Egyptian Government revealed the following minute deviations of the sides of the Great Pyramid from true north–south and east–west alignments: the north side diverges by one twenty-fifth of a degree from true east–west; the south side errs by one thirtieth of a degree; the east side is less than one tenth of a degree from true north; the west side runs about one twenty-fifth of a degree west of true north. All this was achieved with the pyramid base sides which were about 230 m in length. Such a feat of orientation of a monument so large is remarkable.

The next pyramid completed at Giza was King Chephren's. While still an enormous structure, it is slightly smaller than Cheops' monument. Its original height was about 144 m above its base, 3 m less than the height of Cheops'. Its square base originally had sides of length 216 m compared with the base length of approximately 230 m for the Great Pyramid.

Completing the triplet of Giza pyramids is the smallest and latest, that built for King Mycerinus. Its apex was originally about 66 m above its base whose square sides were of approximate length 109 m. Both Mycerinus's and Chephren's pyramids are beautifully oriented north–south and east–west, almost equalling the precision obtained in Cheops's monument.

Now the fundamental point of these accurate alignments is that we are talking about orientations with respect to the earth's *geographic poles*. The magnetic compass was not known to the Egyptians at this time, but such a knowledge, had they possessed it, would not have enabled the pyramid builders to achieve the present alignments. For the magnetic compass points, or tries hard to point, towards the *magnetic* north pole which is about 10° away from the geographic north pole. No, the Egyptians oriented their pyramids so beautifully along the *geographic* north–south direction (i.e. along the direction of the axis about which the earth is turning in its daily rotation) by reference to the sun and the stars. And as a result the east faces of the pyramids of Giza were always oriented perfectly at right angles to the rising sun on the first days of spring and autumn, like supplicants paying twice-yearly obeisance to the sun god Khepri. If he could create beetles out of dung balls, perhaps he could do something with mummified royalty.

Stonehenge

> ... the immemorial gray pillars may serve to remind you of
> the enormous background of Time.... Henry James

About 130 km southwest of London, on Salisbury Plain, stand the magnificent megaliths of Stonehenge (figure 1.2). About 1000 years after the Egyptians had completed the Great Pyramid, thousands of Stone Age Britons were dragging enormous masses of stone across Salisbury Plain and assembling them into a ring structure with remarkable features. As we shall soon see, these ancient Britons were just as concerned as the earlier Egyptians were with orienting their stone monument in a special direction.

Analysis by archaeologists has established that Stonehenge was not completed in one coherent spell of construction. It was, in fact, built in three separate phases (Stonehenge I, II and III), the construction requiring about 300 years for completion. Stonehenge I (figure 1.3)

Figure 1.2: *Stonehenge.*

was a simple structure and yet full of significance from our point of view. The original ring was formed by a ditch which was flanked on its outside and inside by two large chalky banks. The outer bank was about 116 m in diameter and about 2.5 m wide. The inner bank was roughly 98 m in diameter, 6 m wide and perhaps, originally at least, head high. To the northeast the banks and ditch are breached by an entrance which is about 11 m wide. About 30 m beyond the outer bank the Stonehenge I people erected a massive unworked chunk of local sandstone now known as the 'heel stone'. This stone forms the heart of all astronomical theorizing about the purpose of Stonehenge. The heel stone was transported to its present site from Marlborough Downs, about 30 km to the north. It is about 6 m long, about 2.5 m wide and about 2 m thick, weighing roughly 35 000 kg. No doubt originally erected with its long axis vertical, it now tilts about 30° from the vertical towards the Stonehenge circle.

Another feature which has aroused many imaginations is the set of 56 uniformly spaced holes which mark out a circle inside the inner bank (figure 1.3). These are called the 'Aubrey holes' after John Aubrey (1663?) who discovered them upon noticing some small depressions

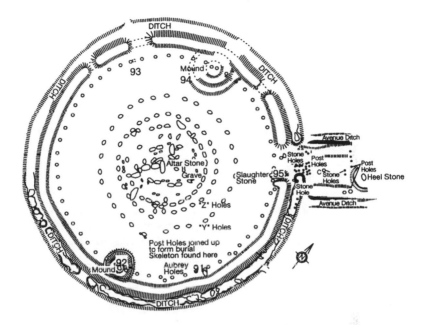

Figure 1.3: *Schematic plan of Stonehenge.*

in the turf. The holes vary in depth between 0.6 and 1.2 m and were filled in with chalk soon after their original excavation. Later, some were re-excavated and refilled with rubble and, sometimes, the cremated remains of humans.

There is some evidence that the builders of Stonehenge also may have outlined a rectangle by using four stones to mark its corners. These are called the 'station stones' and are numbered 91, 92, 93 and 94 in figure 1.3. Only two of these stones (91 and 93) remain. The earlier existence of 92 and 94 is speculated on the basis of two mounds surrounded by ditches which would clearly have served well as mounts for stones. It should be emphasized that it is not known definitely that the station stones formed part of Stonehenge I or whether they were added in the later waves of building. Nevertheless, we shall soon see (pp 9–12) that the orientation of the station stone rectangle has come to play an important role in theories of Stonehenge as an astronomical observatory.

This then appears to have been Stonehenge I—essentially a circular compound bounded by the two mounds, containing the 56 Aubrey holes and a few stones. Outside stood the sentinel heel stone. Few of the enormous stones which we now associate with Stonehenge had yet appeared, but the massive inner mound of white chalk would have impressed most early visitors.

What were the Stone Age Britons up to when they conceived and constructed this curious compound? Why was the entrance sited where it was? Standing in the centre of the enclosure and looking out through the entrance, one could see the heel stone, its tip in line with the distant horizon. The arrangement of the gap in the mounds and the heel stone formed a primitive gunsight. What could Stone Age man have possibly been interested in pinning down on the horizon in his sights? The answer to this question was first provided in 1740 by Dr William Stukeley in his book *Stonehenge, A Temple Restored to the British Druids*. The quarry sought by these early men was, in fact, the midsummer sun.

If you go to Stonehenge next midsummer morning (June 21) and wait for the dawn, then, as you stand at the monument's centre, you will find that the first flash of the sun's rays is slightly to the left of the heel stone. As time passes, the sun raises itself above the horizon while gradually moving to the right. When the sun's disc just clears the horizon, it hovers momentarily exactly above the tip of the heel stone, as though caught in some fantastic balancing act. On other mornings in the year (either before or after midsummer day) the sun rises farther to the right and, in fact, for most of the year the sunrise cannot even be seen from the centre through the entrance! If you stood at the centre of the circle to wait for the rising sun every morning from midwinter on, you would find that it only began to come within your sights (determined by the gap in the bank) as June approached. And only on midsummer morning would it fleetingly hover exactly on target, pivoted on the horizon on the tip of the heel stone. As midsummer passed, and the days grew shorter, the rising sun would fail to reach the target by a gradually increasing amount.

Now virtually all early or primitive societies have observed, recorded and celebrated the four key points in the year, i.e. midsummer day, midwinter day, the first day of spring (the vernal equinox) and the first day of autumn (the autumnal equinox). The heralding of midpoints or beginnings of seasons by these key dates was critically important to both the early farmer and the hunter. It is clear that the people of Stonehenge were no different in this regard and they obviously placed the key emphasis on midsummer day.

At this point we should note that things have changed slightly at Stonehenge over the 4000 years which have elapsed since its conception. Thus in those early days the midsummer sun used to rise slightly farther to the left (the north) than it does today. So by the time that its obliquely rising motion had carried it over the line of sight through the heel stone, there would have been a gap between the bottom of the sun's disc and the tip of the heel stone, as seen by a man about 1.8 m tall standing at the centre of Stonehenge. The gap would be about equal to the diameter of the sun's disc. However, as we noted earlier (and as was pointed out in this connection by G S Hawkins), the heel stone is now tilted towards the circle. If we assume that this giant stone was originally erected vertically, then we find that in Stonehenge I time the sun on midsummer day was, as now, balanced on the tip of the heel stone.

This beautiful midsummer day's arrangement of Stonehenge I is dramatic enough, but there are a number of other features of its orientation which serve to confirm the intent of the neolithic Britons to record important events in the sky. We refer now once again to the station stones, numbers 91, 92, 93 and 94 in figure 1.3.

Firstly, we see that the two short sides of the rectangle defined by the station stones are parallel to the sighting line from the centre of Stonehenge to the heel stone. That is to say, the direction to the midsummer rising sun is precisely recorded by stones 93 and 94 and by stones 92 and 91. But since the midwinter sun sets in exactly the opposite direction, then of course these stones also record this direction. Meanwhile, the two long sides of the station stone rectangle, and the diagonal defined by stones 91 and 93, mark out other very interesting directions. To appreciate this, however, we need to review the motions of the sun and the moon in terms of the cyclical shifting of the directions of sunset and sunrise, and moonset and moonrise.

We often talk loosely about the sun 'rising in the east'. However the sun only rises exactly due east on two days in the year, namely the first days of spring and autumn. From the first day of spring until midsummer day the point on the horizon of sunrise migrates gradually northwards. On midsummer day this northerly trend ceases and reverses itself, so that the sunrise moves southwards until on the first day of autumn the sun rises exactly in the east and sets exactly due west. Between autumn and midwinter day the sunrise moves ever southwards until on midwinter day the motion reverses and by the next spring the sun is again rising due east. This unending to-and-fro motion of the sunrise point along the easterly horizon thus takes place between a northerly extreme and a southerly extreme. Naturally the sunset point is similarly oscillating up and down the westerly horizon between a northerly and a southerly extreme, the trip from one limit to the other requiring 6 months. During a year, then, there are two northerly extreme horizon points (one in the northeast for sunrise, one in the northwest for sunset) and two southerly extreme points. This is clearly seen in figure 1.4. So far so good. What we have just reviewed is basically common knowledge. It was undoubtedly common knowledge in 2000 BC when humans everywhere (who were not living in forests) watched the skies at night out of joy and fear. In contrast, the motion of the moon's rising and setting points is not the common knowledge of today's average high-school graduate. In this regard Stone Age man was more advanced. His nightly observations over the years told him the following. Just as during the year the sunrise and sunset points migrated up and down the horizon, so, during a month, did the moonrise and moonset points oscillate up and down the horizon. But whereas the extreme north and south points reached by the sun were unchanged year after year, the extreme north and south points reached by the moon every month steadily changed from one month to the next! In fact, careful observation at Stonehenge would show that the moon's monthly extreme-north rising point varied between two fixed limits: 29.5° north of east and 50.6° north of east. Similarly, the moon's monthly extreme-south rising point, shuttled steadily between 29.5° south of east and 50.6° south of east. Because of the symmetry of the situation there would be four positions equivalent to these on the westerly horizon at moonset.

If all this seems rather complex, all you need to remember is that the moon is twice as complicated in this aspect of its motions as the sun and so instead of having four key directions in figure 1.4 it has

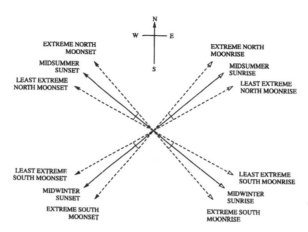

Figure 1.4: *Stonehenge's station stone rectangle encodes six solar–lunar horizon positions (after G S Hawkins).*

eight. Reverting now to our discussion of the directional properties of the station stone rectangle, we find that, whereas the short sides of the rectangle mark out the directions of the midsummer sunrise and the midwinter sunset, the long sides run in the directions of the extreme-south moonrise and the extreme-north moonset (figure 1.5). Furthermore, the diagonal of the rectangle formed by joining station stones 91 and 93 defines the directions of least extreme-south moonrise and least extreme-north moonset (figure 1.5). The station stone rectangle thus has the remarkable property that two of its sides record two of the four principal solar directions; its other two sides mark two of the eight key lunar directions, while one diagonal defines two more of the lunar directions. This remarkable rectangle, then, encodes six of the 12 key solar–lunar horizon positions described in figure 1.4.

The first person to point out the significance of the station stone rectangle was C A Newham, who chose the somewhat unconventional route of publishing his results (on 16 March 1963) in the *Yorkshire Post*, the venerable daily newspaper of the 'county of the broad acres'. Later, Newham summarized his varied thoughts on Stonehenge in a privately printed and circulated monograph entitled *The Enigma of Stonehenge*. Meanwhile, independently of Newham, the British-born astronomer G S Hawkins (of Boston University in the USA) was

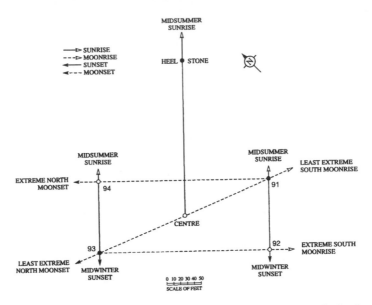

Figure 1.5: *Stonehenge, eight principal directions of the moon on the horizon, corresponding to four of the sun (after G S Hawkins).*

studying Stonehenge alignments and also discovered the interesting features of the station stones arrangement. His first paper describing these results appeared in the widely read British scientific journal *Nature* on 26 October 1963, just 7 months after Newham's newspaper article. Hawkins's note was also interesting because of some other findings that he presented in addition to those connected with the station stones. With the aid of a computer, Hawkins had done the equivalent of laying an enormous ruler along the lines connecting pairs of stones at Stonehenge and noting where the ruler intersected the horizon. He had done this for all the obvious pairs of important stones and thereby derived numerous points on the Stonehenge horizon. Then he had compared these points with the horizon rising and setting points of all the planets and bright stars as they would have been seen when Stonehenge was constructed. Quite unexpectedly he found that the only heavenly bodies with rise and set points (at the time of Stonehenge's construction) agreeing with the Stonehenge alignments were the sun and the moon. Hawkins showed that ten of the 12 key rising and setting points of the sun and the moon (figure 1.4) were recorded by Stonehenge I alignments. Subsequently he found that one of the two remaining points was also recorded so that finally 11 key solar–lunar alignments were found for Stonehenge I. These Stonehenge alignments are so extraordinary that it would be unreasonable to regard them as accidental and it is quite clear that the neolithic Britons built Stonehenge as an astronomical observatory. Exactly why these people built the observatory can, of course, never be known with certainty and speculation on its purpose (particularly by Hawkins) has caused great controversy. However, in the words of C Renfrew, Professor of Archaeology at the University of Southampton, England, Hawkins's 'arguments for at least some of the alignments are now widely accepted, and it is clear that Stonehenge was used to observe the motions of the moon as well as the sun: it was indeed an observatory'.

In the 200–300 years following the completion of Stonehenge I, several periods of construction occurred which brought Stonehenge to its finished form whose remains are visible today. The Stonehenge II workers began the first stone ring (actually a double ring) erected at the site but never completed the installation of the approximately 5000 kg block of bluestone. Their successors, the Stonehenge IIIA people, dismantled the structures contributed by the Stonehenge II folk and introduced about 80 gigantic 'sarsen' stones. It is the residue of these that we usually think of today as comprising Stonehenge. Firstly they built five trilithons and arranged them into a horseshoe shape with the 'mouth' of the horseshoe opening in the celebrated midsummer sunrise direction of the heel stone. By 'trilithons' is meant three stones, in this case two of them vertically upright and one resting athwart them to form an enormous stone archway. Then this trilithon horseshoe was encircled by a set of neatly spaced huge stones (weighing up to about 25 000 kg) which served as supports for approximately 7000 kg blocks which bridged the gaps between the tops of the uprights.

Little remained to be done now. In the Stonehenge IIIB cycle, some erection of bluestones was begun and then suspended and the new stones removed. Two concentric circles of holes (Y and Z in figure 1.3) were then dug for unknown reasons. Finally, the Stonehenge IIIC people installed a series of bluestones at the innermost part of the monument in a roughly semioval outline and set out a circle of bluestones between the sarsen trilithon horseshoe and the sarsen circle (figure 1.3).

These Stonehenge III additions of course produced a greater complexity at the site. Despite this, as Hawkins pointed out in his book *Stonehenge Decoded*, all but five of an original 16 solar–lunar alignments were retained (i.e. not interrupted). In addition, eight of the earlier alignments were duplicated, but this time as views through archways.

We have already noted that these remarkable Stonehenge alignments surely cannot be due to chance and that the early Briton obviously wanted to record the key horizon directions of the sun and moon. In 1964, Hawkins aroused great controversy by suggesting in a paper in *Nature* that Stonehenge was not just an observatory or a calendar, remarkable in itself as that is, but that the great stone collection was really a device for predicting eclipses, i.e. for predicting dramatic events in the future. He argued that it was designed as a Stone Age digital computer. Hawkins noted that it is obviously impossible to prove this but gave arguments to try to show how the neolithic Britons could have used Stonehenge very simply to forecast eclipses. Now there is no denying that the possession of such an ability would have given its owner enormous prestige and power. Eclipses were awe-inspiring events for early man and anyone capable of accurately forecasting temporary extinction of the sun or moon would have been credited with truly magical powers. But there is no reliable record of accurate eclipse prediction in other parts of the world until perhaps the third century BC. Hawkins's claims therefore for eclipse-prediction 1500 years earlier than this in Britain naturally caused a stir.

Hawkins's arguments, in fact, were not convincing. Firstly, the technique of using Stonehenge suggested by Hawkins does not forecast unfailingly eclipses which will be visible at Stonehenge. Thus the 'priests' who would be operating the system would not be infallible and this would be a serious drawback. Doubtless the priests would argue that they had prevented the eclipse which failed to turn up, but Neolithic sceptics might not be impressed. More importantly, Colton and Martin in 1966 pointed out that Hawkins's underlying analysis of the frequency of eclipse occurrences was too simple and that the method of using Stonehenge advocated by Hawkins was unreliable. Meanwhile the famed English astronomer Fred Hoyle had become interested in Stonehenge and had verified the alignments suggested by Hawkins. He also liked the idea of early Britons forecasting eclipses, but he did not like the methods which Hawkins had proposed. He therefore put forward a most ingeniously simple recipe by which the people of Stonehenge could move three counters (small rocks, say) around the circle of 56 Aubrey holes in such a way that when the three counters all arrived simultaneously at one Aubrey hole position an eclipse of the sun or moon would occur. The theory underlying this imaginative scheme is correct. However, it is also striking in that it predicts all eclipses which ever occur. While it is true that many of these eclipses would not be visible (even in ideal weather conditions) from Stonehenge, it is also true that no eclipse would occur at Stonehenge which had not been predicted. The only real flaw in Hoyle's thesis, which in my opinion is a fatal flaw, is that while Hoyle's recipe for moving the counters could be followed by any child who could count, the underlying theory was not available to the Stone Age Britons. In the present state of our knowledge and education the theory is simple and could be successfully explained to most high-school students. But, in the early Britons, such a state of sophistication is too much to accept. It requires an understanding of the relative motions of the sun, moon and earth which came much later in man's history. It is asking too much to believe that people who did not leave a written record had developed an appreciation not only of orbital planes but also of the imaginary line of intersection of two orbital planes rotating in space through 360° every 18.61 years.

Two things are clear, then, about Stonehenge: firstly, it undoubtedly recorded precisely the important horizon positions of the sun and the moon as seen almost 4000 years ago, and this can scarcely have happened by chance; secondly, no good argument has yet been advanced to show that Stonehenge was used as an eclipse predictor, even though Hoyle can certainly use it, even today, for this purpose.

Stonehenge is not the only neolithic stone circle in Great Britain. It is the most striking and elaborate, but many lesser stone circular arrangements are known in the British Isles (and in northern Europe). These have been carefully surveyed by the engineering professor, Alexander Thom, who has shown that the solar–lunar alignments of Stonehenge are found in many of them. As we would expect, the motions of the sun and moon were keenly watched and recorded for many generations. Stonehenge, while by far the best-known primitive observatory and timepiece, was not alone.

The Chinese oracle bones and the evolution of calendars

While the Stone Age Britons were recording the key horizon positions
of the sun and the moon at Stonehenge and numerous other sites,
across the world in China a neolithic tribe began to emerge from its
competitors. Its members were destined to become known to us as
the Shang people and they made enormous strides in the development
of civilization in China. The precise dates of the Shang Dynasty are,
of course, difficult to set, but their period of predominance in China
appears to be roughly 1750–1100 BC. Their importance from our point
of view is that, like the Stonehenge people, they carefully studied the
motions of the sun and the moon. Unlike the Stonehenge folk, however,
the Shang people had learned how to write for posterity. Whereas the
neolithic Britons left their knowledge of the sun's and moon's motions
encoded in arrangements of gigantic stones whose true message was
decoded by Newham, Hawkins and Thom, the Shang people sat down
and wrote about what they observed. Miraculously, it now seems, they
chose sometimes to 'write' on bone, and as a result their observations
survived and may be read today. The Shang scribes told of many things:
of sacrificial ceremonies, of farming matters, of wars, of hunting, of
dreams and of childbirth. Most importantly for us, however, they
described their calendar.

Apart from hunting and tool making, we might suppose that calendar
keeping was man's oldest profession. For a calendar is merely a means
of keeping track of the passage of time, from day to day, month to
month, season to season and year to year. The motions of the sun and
moon were the first features used by early man. The passage of days
was marked by the risings and settings of the sun. Somewhat longer
time intervals were marked by the reappearance of the crescent moon
about every 30 days. The recurrence of a particular season, say winter
in northern lands with its poor hunting and slack farming, provided a
first crude measure of the progress of years. It would not be long before
humans noticed that about 12 crescent moons could be counted between
successive winters. Such observations would not have come from the
joy of intellectual speculation and scientific enquiry but because, by
keeping track of the number of crescent moons, one could plan ahead
for hunting or farming. And those who planned ahead survived; so
calendar keeping had enormous survival value and would have once
been an important factor in the course of human evolution. Calendar

keeping had positive feedback.

Now keeping track of the years by noting the arrival of winter, say, would have been done initially by observing the onset of cooler weather. But this, of course, is not very precise. Confusion would result from exceptionally cool or exceptionally warm late autumns. At some very early stage in his development, man therefore found much more accurate means of fixing the seasons and, accordingly, the years. At Stonehenge he built an observatory in which he could catch the sun on the heel stone on midsummer morning. In many other parts of the world he used the much simpler device of watching the changing length of a shadow cast by a vertical stick.

This was the method adopted by the Shang people. On any clear day of the year the length of the shadow of a vertical stick changes steadily from hour to hour. When the sun is very low in the sky at dawn, the shadow is naturally extremely long. As the morning passes the shadow shortens until, exactly at noon, its length is a minimum. The shadow lengthens again, during the afternoon hours, until it fades from sight at dusk. This simple pattern varies only slightly from day to day, the important variation being that the length of the shortest shadow (i.e. the noontime shadow) changes from one day to the next. At noon on midsummer day the shortest midday shadow appears, while the longest noontime shadow is cast on midwinter day. The Shang people erected a post called a *kuei* and, using its shadow in this way, detected midsummer and midwinter, calling these times *chih* meaning 'limits'. The passage of the years could thus be followed very precisely by the recurrence of the *chih*, and the arrival of spring and autumn could be estimated as being about half way between the *chih*. The Shang people could now count the number of days in a year and record their results on bone.

The Shang folk inscribed various kinds of bone and shell (figure 1.6). Mostly they made use of the nether part of tortoise shells (the so-called plastron). Occasionally animal scapulae, rib bones and skulls were used. Even fragments of human skulls were employed. The tortoise plastrons were favoured, no doubt, because they were wide and flat and relatively easy to prepare. Some were perforated along an edge, evidently for stringing into bundles to form bone books. Stacks of bones excavated from Shang deposits are regarded as China's oldest books.

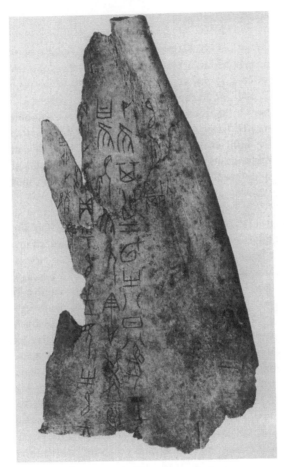

Figure 1.6: *Chinese oracle bone.*

The name 'oracle bones' comes from the practice of the Shang people of addressing enquiries on the bones to their dead ancestors. These early Chinese (like pyramid-building Egyptians) believed that their ancestors enjoyed a life after death and made sacrifices and offerings to them. The supposed replies from these ancestors were also recorded on the oracle bones. The Shang scribes first wrote their messages on bone in ink, using a brush, just like modern Chinese calligraphers. Then, using a sharp-edged jade or bronze tool, they cut out the symbols from the bone. So delicate was the carving that individual writing styles were clearly preserved. The Chinese scholar Tung Tso-pin translated many of these inscriptions on the oracle bones and from them found the number of days in the year in Shang times. Averaging over a period of 152 years he found that the year contained 365.25 days, in excellent agreement with the modern value of 365.2422 days.

However, this is not the only interesting feature of the Shang calendar. To appreciate these new points, however, we must digress briefly to explain some of the difficulties that one runs into in trying, as man has done throughout history, to keep track of time using two clocks, the sun and the moon, which run at utterly different rates.

Suppose that you were a tribesman living in some undeveloped area of the world, and furthermore suppose that you decided to use the moon to record the passage of time—just as the North American Indian was doing at the time of Columbus' explorations. Then, little time would elapse before you became aware of an irritating feature of the lunar clock! You would find that the time passing between identical returns of the crescent moon was not a whole number of days. For the length of the month is not exactly 28 days, or 29 days or 30 days—it is 29.5 days! To be more precise it is 29.530 59 days. Once your irritation had subsided, and if you were ingenious and the least bit mathematical, you would hit on the following easy way of making your 'month' have a whole number of days in it and yet keep in step with the real moon. You would let one month have 29 days, while the next had 30 days, and you would keep up this succession of alternately long and short months, first 29 days, then 30 days, then 29 days, and so on. For in any long time interval many moons would pass and one half would have 29 days and the other half would be 30 days long, so that, taken over a number of months, the average length of the month would be half-way between 29 and 30 days, i.e. 29.5 days. In this remarkably simple way your average month would be within about one tenth of a per cent of the correct modern value. The ancient Greeks used this method about 400 BC, calling the 30 day and 29 day months 'full' and 'hollow', respectively. In fact this approach to regulating the lunar calendar has been followed by many succeeding generations for religious purposes.

Returning now to the oracle bones we can see what Tung Tso-pin discovered about how the Shang thinkers handled this problem. They had in fact coined the two different sorts of months: the *ta-yueh* or 'large month' of 30 days, and the *hsiao-yueh* or 'small month' of 29 days. These were used alternately in succession throughout a 12 month year so that a 29.5 day month, on average, was being used. The oracle bones show, therefore, that the length of the month was virtually the same well before 1000 BC as it is now.

The acuteness with which these early Chinese observed the motions of the moon rivalled that of the earlier Stonehenge people who precisely marked the two northerly and two southerly extremes of moonrise over a cycle of 18.61 years. For Tung Tso-pin found that, in an occasional year, the regular 29–30 day alternation was broken by just once allowing two 30 day months to follow one another at the beginning of the year. By thus having one extra month of 30 days in that year, and therefore one fewer month of 29 days, one was very slightly increasing the average length of the month. To find this mean month, Tung Tso-pin averaged the lengths occurring over a period of 152 years and found a value of 29.531 days. This is a fascinating result, because as we saw a few moments ago, the present value for the length of the month is 29.530 59 days. The oracle bones therefore demonstrate that in Shang times the month differed in length from its present value by no more than about one part in 70 000, which is one seven-hundredth of a per cent.

Before we leave the evidence excavated from the north China plain in the past 40 years there is one final feature of calendar keeping which is significant in any considerations of possible disturbances of the earth–moon system in historical times. This point arises from the simple numerical fact that 12 months of 29.5 days' duration do not amount to a year of 365.25 days. 12×29.5 is actually 354 days! Thus when 12 such months have elapsed, the solar year still has about 11 days $(365 - 354)$ to run. After 3 solar years, then, the lunar calendar will be $3 \times 11 = 33$ days ahead of the solar calendar. That is to say, if you took as your 'year' 12 months of 29.5 days each, then relative to the solar year, which of course determines the arrival of the seasons, you would be wrong by 33 days, i.e. about 1 month, after every 3 years. Thus the first day of spring, for instance, would come later and later in the months on your lunar-based calendar. Instead of coming regularly on 21 March, as it does on our modern well regulated calendar, it would come on 21 March one year, but in late April 3 years later and so on. If therefore you wanted to stick with your lunar calendar, as many societies did for centuries, but you also wanted your seasons and their often-associated religious festivals to recur reasonably close to a fixed date in that calendar, then you would have to resort to a simple regulating mechanism. For example, you could make every third year be 13 months long. In fact, such a device of adding 1 month to a year every so often was used by many ancient societies, the additional month being known as an intercalary month. However, this simple use of the intercalary month every third year is not precise enough, because we found that the lunar calendar is out of step by 33 days after 3 years and not 29.5 days. Therefore what we find is that, rather than a 3 year cycle, a 19 year cycle must be used and it must be used in the following way; 12 of the years in the cycle should contain 12 months each, while the remaining 7 years should have intercalary months and therefore be each of 13 months' duration. Just such a calendar as this, based on a 19 year cycle with seven of these years having an intercalary month, is still the basis of the Jewish calendar and has been for over 2000 years. Many other races have also used it successfully. Perhaps not surprisingly, now that we have seen the degree of sophistication reached by the Shang people over 3000 years ago, we find from Tung Tso-pin that these early Chinese used intercalary months. They called them *jên-yueh* and, as noted by Chêng Tê-k'un (lecturer in Far Eastern Art and Archaeology, University of Cambridge, England), 'Seven intercalary months were added every 19 years.' These illustrations of how the solar and lunar clocks have maintained the same relationship since well before 1000 BC would appear to leave no doubt that the recent cosmic disturbances of the earth proposed by Immanuel Velikovsky in best-selling books such as *Worlds in Collision* have simply not happened.

CHAPTER 2

THE AGE OF THE EARTH;
THE GENESIS BURDEN

The discovery that the earth is 4.5 billion years old is one of the greatest triumphs of twentieth century science. For over 5000 years of intellectual and social evolution the leading thinkers of great civilizations remained in total ignorance of even a crude approximation to the true state of affairs. For the first millennia this was hardly surprising, since there was no obvious meaningful way of measuring such an extraordinary interval of time. The far simpler task of keeping track of time from day to day and from year to year gave sophisticated civilizations fits, as they struggled with intercalary months (see p 21), leap years and the like.

For over 1500 years, the intelligentsia of Europe accepted the romantic Judaeo-Christian Genesis version of the origin of the universe with its equally misleading indications of an age for the earth (universe?). Even the greatest of scientists, Isaac Newton, fell victim to the biblical myths. The most influential early estimate of the age of the earth based on biblical interpretation was produced by Julius Africanus in the third century AD in his *Chronographia*. According to his interpretation of Genesis, the whole of history would be encompassed in a 'cosmic week' of seven days—each day being 1000 years long! In justification for these long days, Julius Africanus appealed to verse 4 of Psalm 90: 'A thousand years in thy sight are but as yesterday.' The first 6 days of the cosmic week were to include all time elapsing between creation and the second coming of Christ—a hypothetical event awaited by many Christians as heralding a new heaven and earth according to the Book of Revelation.

Africanus decided that five and a half of his cosmic days (i.e. 5500 years) had elapsed between creation and the first coming of Christ, so that in his opinion the earth was almost 6000 years old. Not only that, he could obviously predict that Christ would come again in AD 500. This would bring us to the final millennium (the seventh day of the cosmic week), which would represent the 1000 years' Kingdom of the Messiah which would be followed by the end of the world!

Despite the fact that nothing miraculous occurred in AD 500, Africanus' 6000 year age estimate survived into medieval times. The great religious reformer Martin Luther stated that the earth was formed in 4000 BC, an estimate which was refined to 4004 BC by Archbishop Ussher. Such was the grip of the Bible on men's minds that only in the eighteenth century did a few daring thinkers, such as the great philosopher Immanuel Kant and the brilliant scientist, the Compte de Buffon, reject Africanus' imaginary cosmology, and break the 6000 years' mould for the age of the earth. In the *General History of Nature and Theory of the Heavens*, Kant described a staggeringly modern view of the universe which he saw as infinite in space and time. Kant imagined star-filled galaxies condensing out of the background material of the universe. Africanus' 6000 years were ejected into the cosmic rubbish heap. 'There had perhaps flown past a series of millions of years and centuries, before the sphere of ordered Nature, in which we find ourselves, attained to the perfection which is now embodied in it.... Creation is not the work of a moment.... Millions and whole myriads of millions of centuries will flow on, during which always new Worlds and systems of Worlds will be formed...'. After the death of his protector Frederick the Great, Kant suffered the usual fate of those thinkers who challenged the accepted interpretations of the Bible and had to recant.

Meanwhile, Georges Louis Leclerc, Comte de Buffon picked up an idea abandoned by Newton and began to make an actual numerical estimate of the age of the earth. Newton had observed how a variety of hot bodies cool down and, in that laterally thoughtful way of his, had applied the cooling law that he had discovered to the earth. Assuming that the earth were made wholly of iron, and supposing that it might have been at red heat at formation, Newton estimated that it would take 50 000 years for it to cool to its present surface temperature. To one so steeped in biblical lore, this was totally unacceptable and was rejected. He said, 'I should be glad that the true ratio was investigated by experiments.'

Buffon took Newton up on this and performed a series of experiments on cooling spheres of different compositions and sizes. Then he calculated the times required for the various planets to cool from white heat to an inhabitable temperature. Buffon found that after 100 696 years, the earth would have attained its present temperature. He then corrected this to allow for the effect of 'calcareous materials which cool in a shorter time than ferrous materials...', and for the warming effect of the sun, finally ending up with 74 832 years! Life had first appeared on earth 37 849 years earlier when the surface temperature was low enough, estimated Buffon. Life on the moon, he suggested, had been extinct for 2318 years because of its rapid cooling. The Comte's calculations, we now know, were way off the mark but, far more importantly, he had taken the epochal step of both breaking with biblical authority and applying the currently known laws of physics to estimating the earth's age. His calculations foreshadowed those of Lord Kelvin, who ironically was to take up the laws of physics against the geologists in the nineteenth century, motivated both by his brilliant vision of the cosmic significance of the arrow of time and by a vain opposition to Darwin's godless theory of evolution via natural selection.

The nineteenth century, in fact, turned out to be the ultimate battleground in the struggle to remove the Genesis burden from men's minds. For the first time the intimately linked concepts of the vastness of the age of the earth, and biological evolution via natural selection, came to be grasped at last and wondered at!

The removal of the Genesis burden

The final widespread removal of the Genesis burden came about because of the advances made in geology in the last half of the eighteenth century and the first quarter of the nineteenth century. Great progress was made in the recognition of minerals and rock types, and the competing significance of water and heat in rock formation was hotly debated by Neptunists and Plutonists. The self-taught William 'Strata' Smith criss-crossing Britain as a surveyor during the building of the canal system was able to correlate sediments over hundreds of miles through similarities in rock type and resemblances among fossils. Episodes of mass extinctions of many species were found frozen in sedimentary strata. Evidence of much volcanic eruption and mountain building was found in the igneous and metamorphic rocks. And so it dawned on many geologists (although, initially, by no means all) that it was absurd to squeeze all this extraordinary history of turmoil and peaceful, painfully slow sedimentation into a mere 6000 years.

Out of this ferment of observation and field work, two men surfaced holding on high a brand new time scale for the earth. Said James Hutton in a paper published in 1785, 'The result, therefore, of our present enquiry is that we find no vestige of a beginning—no prospect of an end.' It may be that this Scottish gentleman farmer, physician and geologist was the first human being to catch a real glimpse of the truly overwhelming length of geological time. What Hutton so eloquently said was true at that time! Given the geological evidence then available, there was 'no vestige of a beginning—no prospect of an end'.

Influenced by his years of field work in Britain, Hutton had rejected the widespread ideas that the record in the rocks was one of repeated spectacular catastrophes and floods. Instead he painted a picture of continents being slowly destroyed by erosion and being slowly rebuilt by sedimentation, volcanic activity and igneous intrusion. And it was obvious from the minute changes in the earth's landscape caused by erosion and sedimentation in a few thousand years of civilization that such processes must have required untold millions of years to produce the many cycles of building and rebuilding Hutton saw in the rocks.

Hutton's remarkable concept of an extremely old earth of essentially *indeterminate* age was taken up by the English geologist Charles Lyell (1797–1875) and developed in detail in his enormously influential book *Principles of Geology* published in 1830. The Hutton–Lyell concept was called 'uniformitarianism', and during the middle of the nineteenth century it came to be the dominant philosophy in geology, gradually relegating catastrophism to the sidelines. 'The present is the key to the past' was their credo; geological processes going on now, at their current rates, are sufficient to explain the past evolution of the crust of the earth if, of course, one is willing to allow the vast intervals of time that would obviously be needed by such incredibly slow processes as erosion and sedimentation.

Lyell's ideas on time were to have a profound influence on two giants of Victorian science: Charles Darwin and William Thompson, who was to become Lord Kelvin. Each was a genius, but with dramatically differing talents. Darwin seems to have been devoid of any formal mathematical ability and had a very patchy training in the sciences. He was, however, an observer of genius, a man who could step into a South American jungle and immediately read the life-styles of its inhabitants (insect, bird or whatever) as easily as others might read a book but, in addition to this magic eye, Darwin was gifted with an unequalled ability to understand the significance of these life-styles and the consequences of their interactions. When these two talents were wedded to his extreme tenacity of focused thought, Darwin was to become the right man in the right place at the right time to revolutionize our understanding of our place in the universe.

In complete contrast, Kelvin was brilliant in mathematics, physics and engineering. Where Darwin was slow to find his feet, Kelvin had published his first paper, on the mathematical work of Joseph Fourier, at the age of 17. Darwin's life was dominated by one major idea whose consequences he examined minutely from every possible direction. Kelvin leapt from subject to subject with remarkable ease, contributing new ideas in thermodynamics, electromagnetism, dynamics and geology. His numerous engineering patents helped to make him wealthy and his consulting work on the laying of the first trans-Atlantic cable popularized his name. J J Thomson, the man who discovered the electron, said of Kelvin, 'His personality was as remarkable as his scientific achievements, his genius and enthusiasm dominated any scientific discussion at which he was present. He was, I think, at his best at the meeting of Section A (Mathematics and Physics) of the British Association. He made the meeting swing from start to finish, stimulating and encouraging as no one else did, the younger men who crowded to hear him. Never had science a more enthusiastic, stimulating, or indefatigable leader.'

Because he scattered his attentions so widely, however, Kelvin's reputation is not as high as it was in his lifetime, whereas Darwin's reputation is secure. Yet, Kelvin's discovery, together with German physicist Rudolph Clausius, of the second law of thermodynamics and his broadcasting of its universal significance were epochal contributions to physics. The second law remains at the forefront of discussion in modern physics, as popular books by such outstanding modern figures as Ilya Prigogine, Stephen Hawking and Roger Penrose attest.

With two such divergent personalities, it is not surprising that the reactions of Darwin and Kelvin to Lyell's concept of limitless time should have been polar opposites.   Darwin welcomed 'uniformitarianism'; Kelvin rejected it! For Darwin, Lyell's endless time supplied exactly what he needed for his theory of evolution by natural selection. To Kelvin it was anathema. Lyell's image of the earth was essentially that of a perpetual motion machine—and Kelvin's second law of thermodynamics said this was impossible. It showed that all things ran down in time, even if, as in the case of the earth, that time should be measured in millions of years. Lyell's first edition of his *Principles* pre-dated the discovery of the second law of thermodynamics by over 20 years, so it is understandable that he would easily violate its tenets. However, even after the wide broadcasting of its meaning, Lyell still showed no clear sign of grasping its demand of a finite age for the earth. This cavalier treatment of such a fundamental tenet was like waving a red flag in front of Kelvin.

Kelvin's evolution

Kelvin's interest in the age of the earth spanned at least 61 incredibly full years. He first spoke on the topic in 1846 as a 22 year old on his inauguration at the University of Glasgow and returned to the issue many times in later life. Motivating his interest initially was undoubtedly his reading at the age of 16 of eminent French mathematician Joseph Fourier's *Analytical Theory of Heat* and other works in which the problem of the earth's age and cooling were discussed. But after 1852, his predominant concern was to show the geologists, at least the uniformitarians, that the earth must have formed only a *finite number* of years ago, and furthermore that the intensity of geological processes must have been decreasing with time. His first major attack on the problem, in 1854, however, came via the sun, which Kelvin correctly realized also could only have a finite lifetime.

The result of his first calculation of the age of the sun, in 1854, shows just how utterly unprepared many scientists still were for the full measure of the age of the solar system. Reasoning that the sun's energy came from the infall of meteors, Kelvin came up with the incredibly short estimate for the duration of the sun's heat of 'not many times more or less than 32 000 years'. The earth was therefore presumably of this age. Soon after this, the physicist seems to have realized the absurdity of this estimate, and he adopted the approach of Herman von Helmholtz, the German polymath, who suggested that the sun's heat came from the gravitational energy released during its formation. This led Kelvin to the conclusion that the sun was almost certainly less than 500 million years old, and most probably less than 100 million years old. He 'pooh-poohed' Darwin's estimate of 300 million years for the age of the chalk formations in Kent which was based on presumed rates of erosion of the chalk.

The Origin of Species had now been out for 3 years and, of course, had caused a sensation with its concept of evolution and especially of evolution guided by the blind randomness of natural selection. Undoubtedly, Kelvin's interest in the earth's age, always strong, now became one of his obsessions. Clearly huge amounts of time were required by Darwin's proposed evolution via the accumulation of minuscule changes from one generation to another, and if Kelvin could limit this time, he could strike a devastating blow against natural selection. It was not that Kelvin was against evolution *per se*, merely the randomness of Darwin's proposed driving force. For Kelvin believed that the universe displayed an order that could only have resulted from design, the designer, of course, being God.

To reinforce his solar estimate, Kelvin now returned to the topic he had chosen for his Glasgow inaugural lecture 17 years earlier, the history of the earth's cooling. Kelvin assumed that the gravitational energy released on its formation would cause the earth to be molten initially. However, he reasoned that, as the planet radiates heat into space, heat would be carried to the surface by convection currents, so that the earth would gradually solidify from its centre outwards until the whole sphere was a solid hot ball at a uniform temperature. Assuming this value was about 7000 °C, assuming a value for the rate at which heat flows through rocks (Kelvin had actually made measurements on this property of rocks), and assuming a value for the rate at which temperature increased with depth in the earth (which he estimated from measurements in mines), it was then a simple exercise in Fourier's mathematics (for Kelvin) to calculate how long it would take for the earth (ignoring its curvature) to establish its present surface temperature gradient.

Where Newton had found 50 000 years and Buffon 74 832 years for the cooling earth, Kelvin now concluded the earth's crust had probably solidified 98 million years ago. A careful analysis of his assumptions led him to suggest that in fact the age could be anywhere in the range 20 million to 400 million years. Julius Africanus' 6000 years barrier had been broken with a vengeance. On the other hand, Lyell's unlimited aeons had been clearly limited, as far as Kelvin was concerned.

During the next 35 years, he returned to the problem again and again, and as his prestige rose and rose, so his limits on the age of the earth became more and more stringent, and the age of the earth dropped and dropped—from less than 400 million years in 1862 to less than 100 million years, to less than 50 million years, to 20–50 million years, finally bottoming out at nearer 20 million than 40 million years, and close to 24 million years in 1897. This was no doubt satisfying to him, as he had revised his estimate of the sun's age to a maximum of 20 million years. At last his considerations of gravity and heat flow had reconciled the ages of the sun and the earth. An ironic situation, however, had now arisen, beautifully described by Burchfield, where the majority of geologists had come over the years to be comfortable with Kelvin's 100 million years estimate, but now felt it absurd to contemplate an earth a mere 20 million years old. Numerous estimates by geologists (using rates of formation of sediments) in the latter part of the century had scattered around Kelvin's 100 million years value (although some suggested significantly older ages) and this nice round figure had begun to assume a widespread acceptance, making geologists loath to follow Kelvin on his downward age-spiral. The increasingly arrogant manner in which Kelvin, ennobled in 1892, and his cohort P G Tait, pontificated on the subject also irked the geologists, who at long last had begun to receive some support in the debate from other members of the physics fraternity.

The key mathematical physicists to enter the fray were Darwin's second son George, a brilliant geophysicist, John Perry, a former student of Lord Kelvin's, and Oliver Heaviside, an eccentric genius who had never attended a university as a student, but who made seminal contributions to mathematics. These men pointed out clearly the great uncertainties buried in Kelvin's assumptions and showed how changes in these would allow the earth to be far older than even Kelvin's original upper limit of 400 million years. They did not say his estimates were wrong—but it was wrong of him to place so much confidence in his results.

However, both Kelvin and the geologists were to be overtaken by events surrounding an extraordinary discovery made in Paris by the professor of physics at the Museum of Natural History. Henri Becquerel, in 1896, had discovered that a uranium salt was emitting a strange, penetrating radiation—radioactivity had been discovered. This discovery was perhaps the most momentous in the history of physics, leading us, as it did, into the nuclear age and opening up for mankind the prospect of its immediate extinction. It also impacted devastatingly in two quite different ways on Lord Kelvin's estimates of the earth's age. Firstly, it was found that radioactivity generates heat. Since (inevitably) Kelvin had not allowed for this in his thermal history calculations, his model became totally inapplicable to the real earth. For all anyone knew, the earth might even be heating and not cooling as Kelvin had supposed. Secondly, and ironically, the very phenomenon, radioactivity, that struck the death blow to Kelvin's calculations, also gave us the tool that had been sought for two millennia—a clock which would settle the question of the earth's age once and for all! It would even tell us the age of mankind and (approximately) his parting from the other apes. And these discoveries would show that both Kelvin and the geologists were hopelessly wrong in their numerical estimates. For the earth's age is measured not in millions, nor in tens of millions, nor even in hundreds of millions of years. It is measured in thousands of millions of years.

But while the mechanism of the chronometer was discovered in 1896, eight years were to pass before it was realized that radioactivity might be used as a geological clock, and then another half-century was to elapse before the earth's age was eventually read accurately on it!

CHAPTER 3

THE AGE OF RADIOACTIVITY

The radioactive demolition of Kelvin's six decades of argumentation about the age of the earth began in Paris in 1903 when Madame Curie's husband Pierre, and his assistant Laborde, discovered that the radioactive element radium maintained itself at a higher temperature than its surroundings. William Wilson and, a little later, Darwin's son George noted that the heating effect of any radium in the sun (now known to be an unimportant effect) would invalidate Kelvin's solar age estimates. And in that same year, 1903, the Irish geologist John Joly pointed out that Kelvin's values for the age of the earth could be greatly increased if one allowed for the heating due to radioactivity in the earth. Within the next few years radioactivity was shown to be essentially a ubiquitous component of rocks from the earth's crust and R J Strutt (son of the great Victorian physicist Lord Rayleigh, an old friend of Kelvin) made the remarkable point that there must be a much lower concentration of radioactivity in the earth's relatively deep interior than in the crust, or the earth would be molten. It was plain for all to see that the age of the earth should never again be estimated from heat flow calculations.

One of the first to grasp the significance of all this for the age of the earth was a young New Zealand scientist, Ernest Rutherford, who was then a professor of physics at McGill University in Montreal, Canada. Rutherford was destined to discover the atomic nucleus and be acknowledged as the greatest experimental physicist of the twentieth century. At McGill, he was making a series of discoveries about radioactivity, often in collaboration with Frederick Soddy, a brilliant young Oxford-educated chemist. In 1904, the young New Zealander sailed to England to deliver a lecture at the Royal Institution in London on his remarkable results. Near the end he planned to touch on their significance for the age of the earth. In the audience was the great man himself—the 80 year old Lord Kelvin. Recalled Rutherford, 'I came into the room, which was half dark, and presently spotted Lord Kelvin in the audience and realized that I was in for trouble at the last part of the speech dealing with the age of the earth, where my views conflicted with his. To my relief, Kelvin fell fast asleep but, as I came to the important point, I saw the old bird sit up, open an eye and cock a baleful glance at me. Then a sudden inspiration came, and I said Lord Kelvin had limited the age of the earth, provided no new source of heat was discovered. That prophetic utterance refers to what we are now considering tonight, radium! Behold! The old boy beamed upon me.'

Rutherford's most important contribution to the age of the earth debate, however, was not made in Britain where for over half a century they had made this topic peculiarly their own, but later that same epochal year (1904) in St Louis, Missouri, USA at the International Congress of Arts and Sciences. It had been suspected for some time that the production of the gas helium was associated with radioactivity, and Rutherford suggested that the accumulation of such helium in minerals could be used as a clock to measure geological time. His suggestion was the first step in the slow development of radioactive clocks which would eventually reveal the age of the earth (for all who want to see). Meanwhile, Rutherford was invited to deliver the Silliman Lectures at Yale in 1905 and chose as his topic the ages of the earth and the sun. It is in the proceedings of this lecture that we find the first detailed discussion of how to measure the ages of rocks and minerals using radioactivity.

Rutherford's opening sentences at Yale make the concept of radioactive dating transparently clear. With his superb intuition he commented that the uranium–helium method might well fail because of the diffusion of helium out of minerals during their lifetime. However, he noted that it had been proposed that lead was probably the end product of uranium's radioactive decay chain and that, since lead was likely to be much less mobile than the gaseous helium, a uranium–lead method had great potential. Let us consider these revolutionary words at Yale, spoken in September 1905.

'The helium observed in the radioactive minerals is almost certainly due to its production from the radium and other radioactive substances contained therein. If the rate of production of helium from known weights of the different radioelements were experimentally known, it should thus be possible to determine the interval required for the production of the amount of helium observed in radioactive minerals, or, in other words, to determine the age of the mineral. This deduction is based on the assumption that some of the denser and more compact of the radioactive minerals are able to retain indefinitely a large proportion of the helium imprisoned in their mass. In many cases the minerals are not compact, but porous, and under such conditions most of the helium will escape from its mass. Even supposing that some of the helium has been lost from the denser minerals, we should be able to fix with some certainty a minimum limit for the age of the mineral.'

To illustrate how the method might be used, Rutherford took the analytical data of Ramsay and Travers for the uranium and helium concentrations in a rare but uranium-rich mineral fergusonite and, knowing (from his own work) the rate at which uranium and its daughters produce helium, he could immediately calculate the age. The value Rutherford found was 500 million years—the upper limit of Lord Kelvin's various estimates of the age of the sun! The many and varied uncertainties then surrounding radioactivity, however, can be seen when we note that, in his St Louis address in the previous year, Rutherford had estimated the age of the fergusonite, from the same data, as only 40 million years! The swashbuckling physicist made no reference to the earlier calculation.

Having shown how a uranium–helium clock might work, Rutherford now described how a uranium–lead clock could operate. Crediting the Yale chemist, Bertram Boltwood, with the suggestion that lead was a stable end product of uranium decay, he concluded, 'If the production of lead from radium is well established, the percentage of lead in radioactive minerals should be a far more accurate method of deducing the age of the mineral than the calculation based on the volume of helium, for the lead formed in a compact mineral has no possibility of escape.' Rutherford did not actually present any uranium–lead age calculations in his lecture. However, he did suggest to Boltwood a method of doing this, and so the honour of publishing the first uranium–lead ages fell to the American in 1907. His results on uranium minerals from ten localities in North America, Norway and Ceylon ranged in age from 410 to 2200 million years and, despite the large uncertainties in both the experimental work and the uranium decay scheme, they pointed the way to an ever-older earth.

Meanwhile, the scientist who picked up Rutherford's suggested uranium–helium method most quickly and effectively was R J Strutt (later to become the third Baron Rayleigh), Professor of Physics at Imperial College, London. (While Strutt never attained the eminence of his father, Lord Rayleigh, he was a fine physicist who not only made the fundamentally important observation that the concentration of radioactivity in the continental crust must be much higher than it is in the deep interior but also has his name immortalized in the term Rayleigh–Bénard instability in the theory of fluid convection.) By 1910, he had measured the uranium and helium content of many rocks and minerals and published a summary of ages ranging up to 710 million years for a sample of the mineral sphene from Renfrew County in Ontario, Canada. These results were consistent with Boltwood's evidence for a very old earth, particularly when account was taken of the observations of Strutt and others that helium readily leaked out of minerals. Of his dates, Strutt commented, 'These are minimum values, because helium leaks out from the mineral, to what extent it is impossible to say.'

After this incredibly exciting breakthrough in the measurement of geological time, the way ahead proved to be surprisingly bumpy and slow for almost half a century. There were many hazards to be overcome. The uranium–helium method, Rutherford's first geochronological child, never did reach maturity. As he had suspected from the first, helium loss is so rapid from so many minerals that it was capable in general only of recording very low minimum estimates of age. In contrast, the Boltwood–Rutherford uranium–lead clock was to become a spectacular winner in the dating stakes, but not until the 1950s, for several reasons. All these revolved around the concept of isotopes.

Rutherford, Soddy and Boltwood meet the Marx brothers

The idea was Soddy's, conjured up after he had left Rutherford's Canadian group and returned to England. The astute chemist realized that he could explain a number of puzzling features of radioactivity if atoms were a little more complicated than everybody else thought. At that time, it had only just come to be widely accepted that atoms really do exist, that all elements such as copper, gold and carbon, are each agglomerations of characteristic atoms. What Soddy brilliantly suggested was that there may be *different species* of copper atoms, or gold atoms or carbon atoms. Just as the Marx brothers, while all being Marxes, were not identical, so the copper brothers, or *isotopes*, as Soddy called them, are all copper atoms but are not completely identical. The isotopes of copper differ essentially only in mass. Thus one copper isotope has mass 63 (in arbitrary units) while the other has mass 65. Where any fool could immediately tell one Marx brother from another, two different isotopes of a single element are virtually chemically indistinguishable. And therein lay the difficulty for the uranium–lead dating scheme.

Unbeknownst to Boltwood (who developed the method before Soddy suggested isotopes), uranium has two isotopes, namely uranium-238 and uranium-235. And both are radioactive. Therefore, the uranium–lead clock was really *two clocks*: the uranium-238 clock and the uranium-235 clock. So Boltwood and his early successors were actually trying to read the time on a confusingly smeared-out average of two timepieces. But worse was to come. As Boltwood had correctly surmised, uranium does decay radioactively to lead, but we now know there are four isotopes of lead, four leaden Marx brothers as it were. Also, uranium-238 decays only to the lead-206 isotope, whereas uranium-235 eventually becomes lead-207. So Boltwood should have been calculating two uranium–lead ages: one from the uranium-238 and lead-206 contents, and another (hopefully with the same result) from the uranium-235 and lead-207 concentrations in the same mineral. Instead, he had inevitably lumped together the two uranium isotopes as the uranium parent, and the four lead isotopes as the lead daughter. Naturally his answers were wrong.

But a bad joke gets worse in the telling, as the Marx brothers amply illustrated again. Worse was yet to come. Not only were there two clocks being inadvertently merged: the two clocks ran at *spectacularly different rates*. In fact the uranium-235 clock runs about 6.5 times faster than its sibling uranium-238 clock. It was inescapable then that the uranium–lead clock would be misread before the discovery of isotopes. Given this, it might seem that Boltwood's ages should have been a total write-off. What saved the method from obvious failure and rejection was the fact that there was always far more uranium-238 in a mineral than uranium-235 so that, while the clocks were smeared together by Boltwood, one predominated over the other.

There was still one more sting left in the isotopic tale. As we mentioned above, lead has four isotopes: lead-204, lead-206, lead-207 and lead-208. Most minerals trap some lead when they crystallize at formation. Boltwood, who knew of only one type of lead, was unable to distinguish this initial inherited lead from that generated by radioactive decay of the uranium; so he would always find more lead than he should and on this account would find ages which were too old. In some cases, of course, this effect was considerably reduced by dating uranium-rich minerals such as uraninite, where the initial lead would be swamped by the radioactively generated lead.

Obviously, until a device had been invented which could separate and accurately measure the isotopes of uranium and lead, little real progress could be made with the uranium–lead clock. The breakthrough came at the University of Cambridge in 1914, where Rutherford was now resplendent with a large team, enjoying an enormous reputation, having discovered the atomic nucleus while still at Manchester (his first stop after McGill) several years earlier. It was, however, not in Rutherford's laboratory that isotopes were first detected, but in that of his boss, the brilliant J J Thomson who had discovered the electrons that Rutherford had shown orbited 'his' nucleus. Thomson had invented the *mass spectrograph*, later versions of which were to be called the mass spectrometer, which was to revolutionize (very much later) radioactive dating. In 1914 at the Cavendish Laboratory, Thomson with his marvellous machine first detected isotopes and proved that the noble gas neon (that which lights so many signs today) was composed of two isotopes: neon-20 and neon-22.

The development of the mass spectrograph was then taken over by Thomson's student Frederick Aston who systematically set out to work his way through the known elements, looking for all their isotopes. However, it was not until 1927 that Aston discovered three of the four stable isotopes of lead: lead-206, lead-207 and lead-208. Two years later, two American scientists sent lead separated from a radioactive mineral to Aston for mass spectrographic analyses and combined his results with their own uranium measurements to calculate the first ever age (approximately 1 billion years) obtained allowing for the isotopic composition of lead. These same data were used in a totally different way by Rutherford to calculate when uranium was expelled from the star (which he erroneously thought was the sun) in which it was formed by thermonuclear processes. He concluded that this was no more than 3.4 billion years ago and stated, 'It follows that the earth cannot be older than 3.4 billion years—about twice the age of the oldest known radioactive minerals.'

However, so little progress had been made since the birth of Rutherford's brainchild in 1904 that in 1931, when the brilliant earth scientist Arthur Holmes reviewed the state of affairs, he concluded that the age of the earth 'exceeds 1460 million years, is probably not less than 1600 million years and is probably much less than 3000 million years'. Holmes had begun as a physics student of R J Strutt at Imperial College, London, and had done more than any other scientist to keep alive the development of radioactive dating in the difficult years from 1910 to 1930. In the decade after this remarkable display of uncertainty in 1931 by Holmes, a young American experimental physicist named Alfred Nier was to appear on the scene who would dramatically improve Aston's mass spectrograph into a mass spectrometer and, with this, detect uranium-235, measure its proportion relative to uranium-238 accurately and also confirm Aston's discovery of the last remaining lead isotope—lead-204. Working at the University of Minnesota, Nier would also, between 1938 and 1941, make a set of analyses of the lead isotope compositions of galena (lead sulphide) crystals which, in the years to come, Holmes and two other brilliant scientists would use in famous estimates of the age of the earth.

CHAPTER 4

HOW DO YOU DATE AN EARTH?

How do you date a planet? I don't mean any old piece of rock that you've chipped off some ancient-looking rock formation. After all, the planet itself must have been formed long before the surface rocks were emplaced and sculpted into their present form. No—how do you date the formation of the very earth itself?

The hypothetically most definitive approach would be to break up the whole planet into ever finer pieces, grinding and grinding them down and mixing them all up together. Then you (presumably in a laboratory on another planet) would measure the amount of uranium and lead isotopes in the averaged grains. After subtracting the lead the earth had inherited at its birth (so-called primeval lead that was floating around in the pre-solar gas-and-dust cloud), you could immediately calculate how long it took the uranium to generate the remainder of the lead—the so-called radiogenic lead.

If you had a fine laboratory on this other planet, and if you could grind up the earth, all this could be done and, eventually, after many repeated analyses, you could know the age of the earth with an accuracy of about 1% or less. However, we obviously can't grind up the earth like this. We have to use our imagination and knowledge, to invent a more indirect approach. And this was where Alfred Nier's mass spectrometric analyses of galenas became phenomenally important. Nier himself did not go on to use his data to calculate the age of the earth. The war had just broken out and Nier was soon to be swept up into the American atomic bomb development, the Manhattan Project, leaving the most important aspect of the interpretation of his lead isotope results to others.

Three very talented men, in fact, independently seized on Nier's work. Significantly ahead of the others was the Russian geochemist E K Gerling, working at the Radium Institute of the Academy of Sciences of the USSR, who in 1942 had a brilliant three-page paper communicated to the USSR Academy by Member B G Chlopin, entitled 'Age of the earth according to radioactivity data'. The theory behind Gerling's approach was essentially that adopted by all future age-of-the-earth investigators. What he did quite simply was as follows: he took a comment made by Nier, translated it into a mathematical equation, then substituted Nier's data and solved the equation for the age of the earth. I call his approach 'through an hourglass darkly'. Let's see whether we can shed some light on it.

Gerling supposed that when the earth was formed it contained some uranium isotopes (235 and 238) and some lead isotopes (204, 206, 207 and 208). Obviously the lead had a certain *primeval isotopic composition*, i.e. the lead isotopes were present in certain proportions relative to each other. Now, because of the radioactive decay of the uranium isotopes to lead-207 and lead-206, the isotopic composition of the earth's lead will be steadily changing with time! The uranium–lead system is therefore a beautiful clock in the earth's deep interior, each tick corresponding to a change in the lead isotope composition. And Gerling knew, from the half-lives of the uranium isotopes, the rate at which the lead isotopic composition changes or the rate at which the clock ticked. The problem was: how could one peer inside the earth to read this clock? Turning now to Nier's galenas, Gerling said that galenas are pure lead sulphide crystals and contain *no* uranium. That is, when the galena ore was formed, somehow lead was taken from the earth's interior, *removed totally from its associated uranium* and has subsequently been lying around in an ore deposit like a broken clock, with its lead isotope composition totally frozen into the value that it had at the time of ore formation. That is, galena lead isotope compositions are snapshots of the earth's uranium–lead clock face, taken at different times during the earth's history.

So, if Gerling had known the initial reading (i.e. the primeval lead isotope composition) on the earth's clock face when it condensed out of the solar nebula, he could take this away from the reading on one of his galena clocks and that difference in reading would have been the time which had elapsed between the earth's formation and the time of the galena's formation. If this galena had been collected from a modern deposit formed, geologically speaking, 'today', then obviously this difference in clock readings would be the age of the earth—the difference between the clock readings at the earth's formation and today! Gerling's was a staggeringly beautiful and simple theory.

None of Nier's galenas was, however, of absolutely modern age; so Gerling took the average galena clock readings on the seven most recently formed galenas, which he estimated on stratigraphic grounds were on average 130 million years old. That is their average uranium–lead clock face reading was photographed 130 million years ago. Now came an even more difficult decision: what was the initial reading on the clock face at the earth's formation? Gerling, of course, couldn't know this. So he did the next best thing. He looked among Nier's data for the earliest clock face reading, i.e. the most primitive lead isotope composition with least lead-207 and lead-206, and used this reading as his approximation to time zero. This reading was provided by Nier's data for galena from a deposit at Ivigtut in southeast Greenland. He then found the difference between this clock reading and his reading of the clock at almost modern times and found 3.1 billion years. But his 'modern times' were, as we've seen, about 130 million years ago; so he added this to the 3.1 billion years and found 'for the age of the earth at this day a value of 3.23 billion years (3.23 billion years = 3.1 billion + 130 million years), which we should regard as a minimum'.

However, Gerling also noticed that in Nier's tables there lay another very beautiful snapshot, one apparently taken from a galena deposit near Great Bear Lake in northern Canada that Nier had shown was about 1.25 billion years old. Gerling therefore subtracted the clock face reading on the Ivigtut sample from that on the Great Bear clock and found a difference of 2.7 billion years. That is, the Ivigtut clock was photographed 2.7 billion years before Great Bear's snapshot. But Great Bear was photographed 1.25 billion years ago; therefore, Gerling concluded that the Ivigtut snapshot was taken 2.7 + 1.25 billion years ago, i.e. 3.95 billion years ago!

Gerling, of course, fully realized that, rather than dating the earth, he had really calculated when the Ivigtut galena clock face was frozen and photographed, at either 3.23 or 3.95 billion years ago. But he obviously had at least found a minimum age for the earth, for the earth is surely older than Ivigtut's galena! And so Gerling concluded his remarkable paper with the words, 'From these computations the age of the earth is not under 3–4 billion years.' Gerling was quite right, but his approach and result languished in obscurity because the year was 1942 and the Second World War raged on.

Less than a year after the end of the war, the great geochronological pioneer whom we have already met, Arthur Holmes, now Professor of Geology at the University of Edinburgh, turned to Nier's galena data and used them to estimate the earth's age. It was 35 years since he had written his first book *The Age of the Earth*. The theory underlying Holmes' analysis was identical with Gerling's, but Holmes wrote in complete ignorance of the Russian's brilliant work, published as it was in a Russian wartime journal. And where Gerling had used only nine of Nier's galena analyses, Holmes used, or initially attempted to use, all 25. He began his celebrated paper in May 1946 in *Nature* with the words, 'Ever since the publication by Nier and his co-workers of the relative abundances of the isotopes in 25 samples of lead from common lead minerals of various geological ages, I have entertained the hope that from these precise data it might be possible to fathom the depths of geological time. The calculations are, however, somewhat formidable, and a systematic investigation became possible only recently, with the acquisition of a calculating machine, for which grateful acknowledgement is made to the Moray Endowment Research Fund of the University of Edinburgh. The results have fully justified expectation and indicate that the age of the earth, reckoned from the time when radiogenic lead first began to accumulate in earth-materials, is of the order 3000 million years.' Holmes then went on to present an incredibly tortuous way of extracting the earth's age from Nier's data. Perhaps subconsciously he was justifying the expenditure on his new calculating machine. Regardless, however, of the complexity of Holmes' algebra, his extraordinary paper very definitely showed how lead isotopes would be the basis for our ultimate calculation of the age of the earth.

Meanwhile, the well known Austrian physicist Fritz Houtermans had also realized that Nier's data were a gold mine. Houtermans had become famous because of a paper written in 1927 with a young Englishman, Geoffrey Atkinson, in which they proposed that the energy radiated by stars is generated by thermonuclear fusion of light elements.

Like Holmes, he used the same fundamental equation as Gerling but found a third way of using Nier's data to solve the equation for the earth's age! Where Gerling used the data from nine of Nier's galenas, and Holmes had used all 25, Houtermans in his very brief note published in 1946 in *Naturwissenschaften* used only three galenas to estimate the earth's age as 2.9 billion years with a possible error of 0.3 billion years. At the end of Houtermans' note appears the following: 'Addendum: After the above report had been set in type... I learned of a recent publication by A Holmes in which the quantitative conclusions of the present work were again reached on the basis of Nier's measurements by the use of the relation contained in equation (1).' Like Holmes, however, he was totally unaware of Gerling's earlier work. Houtermans had emigrated to the Soviet Union in 1933 on Hitler's accession to power, only to be arrested in 1937 on suspicion of espionage. Robert Jungk recounts that part of his interrogation by the Soviet secret police involved a 72 hours' beating during which all his teeth were knocked out! In 1940, he was handed over to the Gestapo and spent the rest of the war working on the abortive German atomic bomb project, which he is said to have done his best to delay.

The impact of meteorites

By 1946, then, it had become clear that the earth was at least 3 billion years old, and Gerling, at least, suspected that it might be over 4 billion years old. Gerling was particularly aware that his calculated result gave clearly a *minimum* age for the earth, because he knew only too well that in his search for the reading on the earth's uranium–lead clockface at the time of the earth's formation, he had had to choose the reading captured by the Ivigtut galena. Yet this galena had no doubt formed long after the earth and its reading could not correspond to that of the embryonic earth's clock. To refine his calculation, one could obviously look for ever older galenas with ever older photographs of the earth's uranium–lead clock readings. But there was in fact a much more subtle and superior approach, and it was suggested independently by Houtermans and the American geochemist Harrison Brown of the California Institute of Technology in 1947. They pointed out that iron meteorites contain virtually no uranium. Therefore, the lead that they contain has remained always the same—unaltered since the meteorite crystallized. If, therefore, the iron meteorites and the earth formed at the same time and captured lead of the same isotopic composition, here was Gerling's missing snapshot—the lead isotopic composition of the earth at its formation, i.e. the snapshot of the reading of the earth's uranium–lead clock at real time zero! All you had to do, then, to measure the age of the earth accurately was to measure the isotopic composition of the lead in iron meteorites (i.e. to read the zero-age clockface), to measure accurately the isotopic composition of lead in modern-time galenas (i.e. to read the modern-time clock face reading) and to find the difference between the two clock readings. That is, you would do exactly what Gerling did, only you would use iron meteorite lead instead of Ivigtut galena.

The first scientists to exploit this combined meteorite–earth clock were the American geochemists Claire Patterson, George Tilton and Mark Inghram, and Houtermans again in 1953. The Americans had painstakingly developed the elaborate meticulous analytical chemical methods required to separate and concentrate in uncontaminated form the minute amounts of primeval lead from iron meteorites. They had also constructed a high sensitivity mass spectrometer with which they could measure the isotopic composition of this pristine lead. From the Canyon Diablo iron meteorite, the one which blasted out Meteor Crater in Winslow, Arizona, Patterson and his co-workers found the zero-age reading on the uranium–lead clock, the primeval lead isotope composition. Patterson then presented age-of-the-earth calculations at a conference in Pennsylvania. Having just got the zero-age Canyon Diablo clock reading he had also found the modern-day reading from lead-isotope analyses of lead from oceanic sediments in one case and 'young' volcanic material in another. Subtracting the iron meteorite clock reading from that in the oceanic sediments gave an age of 4.51 billion years, while using the young volcanic lead instead of the sediments yielded an age for the earth of 4.56 billion years. Gerling's intuition was verified—the earth was over 4 billion years old!

Several months after Patterson orally presented his results in America, a paper by Houtermans appeared in the Italian science journal *Nuovo Cimento* entitled 'Determination of the age of the earth from the isotopic composition of meteoritic lead'. Here Houtermans took the zero-age iron meteorite lead isotope clock reading (that Patterson and his co-workers had published separately from Patterson's orally presented full age-of-the-earth calculations) and subtracted it from present-day lead isotope clock readings found on very young galenas, to find 4.5 billion years for the earth's age, with a possible error of 0.3 billion years. These pioneering calculations by Patterson and Houtermans, in linking the meteorites and the earth into one clock, heralded a new era of understanding and suggested strongly that the earth might well be about 4.5 billion years old.

In the next few years, Patterson applied his sophisticated analytical techniques to *stone* meteorites which were presumed to have been formed at the same time as the iron meteorites and the earth and to have trapped the same initial primeval lead. But, unlike the iron meteorites, the stone meteorites trapped significant amounts of uranium and therefore their lead isotope ratios varied in time owing to the radioactive production of lead-206 and lead-207. So, just like the lead in the earth, the changing lead isotope composition of the *stone* meteorites acted as a clock which has run until today. Therefore, by analysing the lead isotope composition in today's *stone* meteorites, i.e. by reading the clock face today, and subtracting the *iron* meteorite clock reading, Patterson could date the stone meteorites. In 1956 he found an age of 4.55 billion years with a possible error of 0.07 billion years for stone meteorites. In this same paper he read the lead isotope clock face in modern ocean sediments and found that it too differed from the iron meteorite reading by 4.55 billion years. So Patterson concluded that the meteorites and the earth were all 4.55 ± 0.07 billion years.

Patterson's brilliant analytical work was immediately widely acknowledged and his value of 4.5 billion years for the age of the earth has become effectively universally accepted. In a sense, the extraordinary two-millennia quest for our understanding of the age of the earth is over. Unless there is something wrong with our understanding of the radioactive process and of the meaning of time itself, that's it—the earth is about 4.5 billion years old. Julius Africanus, Martin Luther, Archbishop Ussher, Newton, Buffon, Hutton, Lyell, Darwin, Kelvin, etc, were all wrong one way or another. Some of course, were far more dramatically wrong than others. But thanks to the serendipitous discovery of radioactivity by Becquerel, and half a century's struggle with experiment and theory, we now have fulfilled Holmes' hope expressed in his epochal 1946 paper that from these efforts it would be 'possible to fathom the depths of geological time'.

Footnote

Of course, Patterson's early measurements have been supplemented by those of many other workers, all using iron meteorite lead for the zero-age clock reading and mostly using galenas for their snapshots of the ever-running terrestrial uranium–lead clock. New measurements have also been performed on the half-lives of uranium-238 and uranium-235, and on the relative abundances of the uranium isotopes. In his book *The Age of the Earth*, Brent Dalrymple notes that the revisions of these factors would shift Patterson's value only slightly, from 4.55 to 4.48 billion years. Thus the newer, more accurate measurements on iron meteorite and terrestrial leads have all confirmed that the earth's age and that of the meteorites is essentially 4.5 billion years. Totally independent dating by completely different radioactive systems fully supports this age for the meteorites. The pyramid builders, the Stonehenge people and the Chinese oracle bone carvers would, I'm sure, be greatly impressed. Newton, 'Thou shouldst be living at this hour!'

CHAPTER 5

MODERN-DAY ADHERENTS OF JULIUS AFRICANUS

We saw in chapter 2 how it required one and a half millennia to rid the majority of minds of the Genesis burden. Even then a further one and a half centuries were needed to measure accurately the age of the earth at about 4.5 billion years. Interestingly, however, small numbers of people who subscribe to various fundamentalist religious dogmas even today remain wedded to Julius Africanus' concept that the earth is about 6000 years old. This is maintained without a shred of significant supporting evidence, in the face of an overwhelming amount of observational material opposed to such an archaic view, and with no scientific theory of any kind to support this belief. Such a belief would presumably be harmless if held by individuals as a strictly personal creed. It becomes potentially dangerous when fundamentalists seek, as many of them do, particularly in the USA, to get equal time for their beliefs to be taught as science in competition with the Darwin–Wallace theory of evolution and its requirement of the enormous time intervals provided by the radioactive clocks.

The promulgation of such arbitrary beliefs, of course, has no place in science classes in schools. The proper place for their examination is in classes on sociology and religious history. Obviously, the enormous time intervals revealed by the radioactive clocks are hammer-blows against fundamentalist dogma; so it is not surprising that a fundamentalist named R V Gentry, a trained scientist, would try to challenge the radioactive clock.

This is a daunting task, given the remarkable agreement which has been found over the past 30 years among the key clocks: the uranium-lead, rubidium–strontium, potassium–argon and samarium–neodymium

46

systems. In order to shoot them down it is hopeless to attack each individually. Obviously one would rather try to show that there is something wrong with our concepts of radioactivity and its rate of decay as measured by the various half-lives. If one could strike in this way at the very conceptual heart of all the dating methods simultaneously then, if successful, one would have overturned all the various methods and their results in one brilliant coup. Many a scientist (non-fundamentalist) would love to do this, for it would require a revolution in our understanding of time at least as great as Einstein's breakthroughs in special and general relativity theory. True scientists dream of producing such paradigm shifts.

Unfortunately, there is no evidence whatsoever on which to base such a serious challenge. Since modern dating methods work so well, and are so consistent with each other, Gentry had to ignore them completely and look into the early history of radioactive dating, hoping to find a weapon among the long-discarded debris, the tentative results of the early dating pioneers. What he picked out were the interesting observations on pleochroic haloes in minerals made over half a century ago by G H Henderson, a Canadian geologist from Dalhousie University in Nova Scotia.

Such haloes are usually seen with a microscope as small dark circular spots surrounded by dark or coloured rings in mineral thin sections (figure 5.1). They had been known for about a quarter of a century before the Irish geologist J Joly explained in 1907 that they were produced by radioactivity. Joly proposed that the tiny mineral inclusions frequently seen at the centres of the haloes were radioactive and that the passage of particles from the inclusion through the host crystal produced the discolouration seen. Uranium was known to decay through a whole series of radioactive daughters, a number of which also ejected alpha particles. The energies of these were characteristic of each emitter, and so each alpha-emitting radioactive element in the decay chain ejected alpha particles with a characteristic range in a given mineral. So, said Joly, the various rings in a halo corresponded to the ranges in the mineral of the α particles from particular members of the uranium decay chain. Herein was a beautiful, almost photographic, record of the radioactive process, as it had gone on for hundreds of millions of years. Joly's explanation was soon accepted and, during the next 30 or so years, a number of investigators, including Joly,

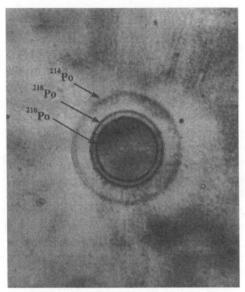

Figure 5.1: *Computer rendering of Henderson's type-C polonium halo.*

examined haloes in some detail. The hope that they might be useful for the measurement of the ages of rocks was a strong motivation. The idea was that laboratory-induced discolouration of minerals by known alpha doses could be compared with the natural discolouration in haloes, thereby indicating the natural dose involved in the generation of the haloes. This natural dose could then easily be converted into an age, if the uranium content of the radioactive inclusion at the centre of the halo were known. These hopes of using pleochroic haloes foundered, however, on the great difficulty of quantitatively measuring the degree of discolouration of a mineral. As a timepiece, the pleochroic halo was an even bigger disappointment than its contemporary, uranium–helium dating.

The study of pleochroic haloes essentially came to an end with the outbreak of the Second World War. Its ending as a subject of research, however, was also marked by the appearance in the *Proceedings of the Royal Society of London* of two interesting papers by G H Henderson (one co-authored by F W Sparks). By 1939, Henderson had examined many mica minerals from around the world and had noted the occurrence of haloes in roughly half of these. In his paper with Sparks,

Henderson set out to classify these haloes into distinct types. Clearly, the commonest was that produced by the uranium family, which was characterized by a dark central disc surrounded by a set of rings with radii corresponding to the alpha ranges of the daughter radioactivities in the uranium decay chain. A variety of other types occurred, however, and we can best explain Gentry's attack on the current concept of radioactivity by looking at one of these: Henderson's type-C halo. This had a dark central disc, surrounded by two rings. The remarkable feature that Henderson and Sparks noticed about these circles was that their radii corresponded to the different ranges of the alpha particles emitted by three isotopes of polonium (an element proudly named by its discoverer Madame Curie after her native Poland) occurring in the uranium-238 chain. Henderson and Sparks therefore concluded that the type-C haloes were due to polonium isotopes. But drawing this conclusion required some hard thinking, because the polonium isotopes all have very short half-lives. For example, polonium-218 from which the other two isotopes are derived by radioactive decay has a half-life of only about 3 minutes. So in 1 hour any given number of polonium-218 atoms will have decayed a millionfold. In 2 hours the initial number will have decayed a trillionfold. Clearly then, if there is any measurable concentration of polonium isotopes (such as would be needed to generate type-C haloes) at a point in the earth, it must have been *produced there* by decay of the much longer-lived earlier members of the uranium decay chain. Thus the polonium rings in the halo should have been accompanied by the rings usually associated with these precursor radioactivities in the decay chain of uranium. In other types of halo they *were* there clear for all to see. Yet here, in the type-C haloes, exactly where the precursor's rings should have been found they were totally absent! It was as though an archaeologist had come upon a large group of healthy, well fed babies in a jungle village with no parents in sight. The archaeologist could have cried, 'My God! These are parentless children. It's a miracle. It shows that our understanding of human beings in general, and their creation in particular, and the way that children are nourished, etc, is all wrong. We must revise our whole understanding of human biology!' Or, she could have exclaimed, 'I wonder where all the parents have gone. Maybe they heard us coming and are hiding in ambush in the jungle somewhere nearby.'

In explaining the type-C haloes in his second paper, Henderson not surprisingly took the archaeologist's second approach. The polonium isotopes, whose traces we find (the halo rings), no doubt had parents, but the parents evidently didn't hang around to produce their own rings. What obviously happened was that, at some stage, fluid flowed through cracks and crannies in the rocks and minerals (a known phenomenon). And this fluid carried dissolved in it many elements, but especially uranium accompanied by the numerous daughters in its chain, such as polonium-218. Henderson supposed that the chemical conditions were just right at some point in a fluid-filled narrow channel in a mica such that polonium would precipitate out of solution there, but no other elements (especially the parents) would. (This is analogous to the principle behind the use of ion exchange columns in thousands of laboratories daily.) Then obviously the polonium being steadily brought up to this point by the fluid flow would keep precipitating there and would produce its halo rings by radioactive decay. *The polonium didn't appear at the point by magic.* It was always being generated in the fluid by radioactive decay of its parents and grandparent. It's just that the parents stayed in the fluid and floated by to somewhere else in the mineral, or out of the rock altogether! This was a beautiful and simple explanation of those polonium haloes found in micas along obvious microconduits in the minerals.

The only potential difficulty with this explanation was that many type-C haloes did not seem to be lying along obvious cracks in the micas. So Henderson said that this must mean that fluids can penetrate *uncracked* micas. On the atomic scale, micas are just like books, being made up of hundreds of *sheets* of atoms piled one on top of another. This sheet-like structure is responsible for mica crystals splitting beautifully along these so-called cleavage planes. And Henderson suggested that evidently, under certain conditions, fluids can in fact flow through *uncracked* micas along these cleavage planes. So far as I know, there was no other evidence for this in 1939 when Henderson wrote.

In complete contrast with Henderson, Gentry looked at these data, threw up his hands and leapt to the archaeologist's first explanation! He couldn't accept that obviously the parents of the polonium were there with the polonium in the fluid and that they simply refused to bail out of the fluid after their precipitating children, and that's why the parental rings are not found in the type-C haloes with the polonium rings. In essence Gentry's position was: It's a miracle. It shows that our understanding of time in general, and radioactive half-lives in particular and their use in dating the earth is all wrong. We must totally revise our whole understanding of time, of the evolution of the earth, of human beings and, inescapably, of the creation of the universe itself.

This of course is just as ludicrous as would have been our archaeologist's behaviour if she had tried to promote a revolution in biology before she searched the surrounding jungle for the missing parents. Gentry's key reason for discussing Henderson's fluid flow hypothesis was that while it would explain the type-C haloes found along obvious conduits in the micas, it would not explain those found in uncracked micas. He considered fluids could not flow through such solid crystals, as Henderson had proposed. However, years later, in 1988, a fascinating observation was published (not at all in connection with polonium haloes) in the leading science journal *Nature* by E S Ilton and D R Veblen, who reported finding *microdeposits of pure copper atoms in the cleavage planes between the sheets in uncracked micas* in granitic rocks surrounding economically valuable copper ore bodies in the southern USA. Presumably, when the copper-bearing fluids penetrated the earth's crust and precipitated the major copper deposits, off-shoots of the liquid flowed into the surrounding granitic rocks, most rapidly into cracks and faults, but apparently many uncracked micas were invaded by the liquid and, at particular points along cleavage planes, conditions were just right to precipitate pure copper into the microdeposits found subsequently by Ilton and Veblen. The mechanism (as was first pointed out by myself and graduate student Hironobu Hyodo in 1993) would be an exact replica of that proposed by Henderson for the point precipitation of polonium (instead of copper) in micas. So much for Gentry's hang-up about fluid flow through uncracked micas!

To maintain that the earth is about 6000 years old in the face of the overwhelming mass of consistent data to the contrary amassed in the last 35 years, and to try to justify this by seizing on misinterpretations of ancient isolated observations, is to bury one's head ostrich-like in the sands of geological time. As I said earlier, it would be wonderful to prove that the earth is 6000 years old. It would herald a breathtaking revolution for the whole of science. Unfortunately there is literally, currently, no evidence on which to base such a revolution.

CHAPTER 6

A CARBON TIME MACHINE

It was 4 billion and 4 BC as Xylon jockeyed his vehicle into a lower orbit, the slightly increased velocity combating the planet's extra gravitational pull. He increased the magnification on the televiewer and peered intently at the screen. 'Can you see anything, Xaro?' he telepathed into the co-pilot's brain. 'Nothing living,' replied Xaro in the same manner, 'Mostly water. Lots of islands.' And with that, Xylon fired up the rockets and headed back towards the mother ship coasting in its much higher orbit.

But Xylon was wrong—for all that he came to planet earth from a solar system where life was billions of years in advance of our own—for 4 billion years ago the earth was teeming with life. Or at least that's what geochemists like Manfred Schidlowski of the Max Planck Institute, Mainz, Germany, say.

The German scientist has been using a carbon 'time machine' to go back to the dawn of earth and detect the tell-tale spoor of untold trillions of photosynthesizing bacteria. For the life then on earth, of course, was not human, or any kind of animal, or plant or fungus. It was microbial, in the form of single cells. Our distant ancestors were bacteria, floating around in great colonies just below the surface of shallow seas, clinging together in huge microbial mats. Some of their descendants clung to this life-style for 4 billion years, and 'living rugs' can still be found today floating in shallow waters in places such as Shark Bay on the west coast of Australia. Luckily for us, other microbial descendants got up off the mat and rode the magic carpet of evolution to the multicellular highly differentiated life which coats the earth's surface today.

How did Schidlowski go backwards in time? How did he flick backwards through the pages of earth's diary, hurrying past *Homo sapiens*, *Australopithecus afarensis* (Lucy), dinosaurs, early mammals, primitive reptiles, amphibians, fish and spineless blobs, in rapid succession? How did he keep on going past page after page of virtually the same pictures, of scarcely changing single eukaryotic cells (whose genetic material, deoxyribonucleic acid (DNA), is kept inside a little bag called a nucleus)?

The answer to these questions lies in the time machine used by the intrepid scientist who has of course yet to convince everyone of the meaning of what he saw. The type of time machine that he used is far and away the best such device developed by man. It is, of course, the mass spectrometer, the very machine that we met in chapter 3, which was invented by J J Thomson (discoverer of the electron) and developed greatly by F W Aston and A O Nier. In today's world of silicon chips and computers it has become an incredibly sophisticated and sensitive device.

In essence it is very simple. Ignoring all the fancy peripherals, it is merely a small tube pumped to a very high vacuum. Inside the tube are maintained electric and magnetic fields. When carbon atoms (all the other elements can be similarly analysed) are admitted as a gas into the tube, the net result is that these force fields create twin orbital streams of two types of carbon atom. Chemically, the two types of carbon atom are for most purposes identical, but the relative sizes of their orbits in the mass spectrometer tell us that one type of carbon atom is 12 times heavier than a hydrogen atom and is called the carbon-12 isotope. The other is 13 times as heavy as hydrogen and is called carbon-13. (Carbon-14 is also an isotope of carbon but, being radioactive and having a short half-life, it is not found in ancient rocks and is not useful for studying the earth's early history.)

During the past 20 years, Schidlowski and many other scientists have studied with their mass spectrometers the isotopic composition of carbon from many sources. To a very good first approximation it seems to be constant; about 99% of carbon is carbon-12 and roughly 1% is carbon-13. On closer examination with high-precision mass spectrometers, however, there prove to be minute, but extremely important variations. The relative amounts of carbon-12 and carbon-13 can vary depending on the history undergone by the carbon.

It turns out that life forms (including you and me) very very slightly, but distinctly, prefer to build themselves out of carbon-12 atoms. Life therefore causes a measurable fractionation of the carbon isotopes at the earth's surface. Many mass spectrometer measurements have shown that the key fractionating feature in life's growth and proliferation from the inert atoms surrounding it is photosynthesis. All the earth's green plants use this process to increase their bulk. In an extraordinarily complex series of chemical reactions, plants extract the carbon that they need from the carbon dioxide in the atmosphere. With the aid of the energy in sunlight trapped by chlorophyll, the plants use carbon dioxide and water to make glucose and many other complex organic molecules, giving off oxygen to the atmosphere in the process. As a result of this sequence, the plants very slightly preferentially extract the lighter carbon-12 isotope from the atmospheric carbon dioxide and therefore leave the inert background slightly richer in the heavier isotope carbon-13.

Plant-eating creatures, animals who eat plant-eating creatures, and animals like us who are both are, of course, all consequently built out of this fractionally lighter carbon. The extraordinary outcome is that life, principally through photosynthesis, controls the relative distribution of the carbon-12 and carbon-13 isotopes at the earth's surface today. In fact, thousands of mass spectrometry measurements have shown that living carbon, on average, is about 26 (arbitrary) mass units lighter than the inorganic carbon in atmospheric carbon dioxide and oceanic bicarbonate.

Now Schidlowski's method of time travel becomes clear. If we could find carbon that had been 'alive' in the past (organic carbon) and compare it mass spectrometrically with equally old never-living (inorganic) carbon we could see whether the tell-tale 26 mass unit difference occurred between organic and inorganic carbon in the past. And if so, we could argue that, since today that difference results from photosynthesizing organisms, then presumably it did so then!

To sample the living and inert past we use the earth's sedimentary rocks. When organisms die, their light carbon eventually gets taken up by sediments, which over time are converted into rock. Tests show that the light carbon preserves its isotopic signature during this process and also as it is gathered into a complex insoluble organic material called kerogen.

To analyse the past's living carbon, therefore, we extract the organic carbon from the kerogen in ancient sedimentary rocks. In contrast, the earth's inorganic carbon is precipitated from the ocean in carbonate rocks. The past's inorganic carbon is therefore collected from old carbonate sediments and mass spectrometrically examined.

Schidlowski compiled over 10 000 carbon isotope measurements (by him and many others) made on sedimentary rocks increasing in age (and therefore going back in time) to 3.5 billion years. While there was considerable scatter, he claimed that, on average the 26 mass units difference between organic and inorganic carbon holds up through time. (The reader should examine the data summarized in figure 6.1 to see whether he or she accepts the argument.)

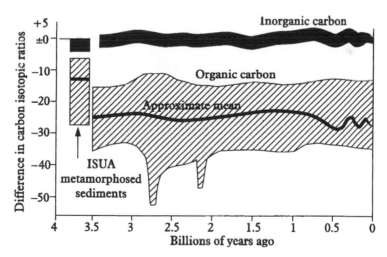

Figure 6.1: *Carbon time machine results (after M Schidlowski).*

To catch a glimpse of an even more distant past, Schidlowski looked at the data from the oldest sediments known—3.8 billion-year-old metamorphosed rocks from Isua, Greenland. These showed a mass difference of only 13 mass units. But Schidlowski pointed out that metamorphism tends to reduce the carbon mass difference and that the Isua data are at least consistent with the rest. Schidlowski said that this overall picture suggests that photosynthesis 'emerged very early in the earth's history, with biological processes consequently leaving their mark on the terrestrial carbon cycle as early as almost 4 billion years ago'.

One more feature of the carbon isotope results struck the German scientist. For the photosynthetic process to produce a 26 mass unit difference between organic and inorganic carbon, it is necessary that about 20% of the total carbon available at the earth's surface be organic. But since this applied 3.5 billion years ago just as much as today (since the 26 unit mass difference has remained roughly constant all this time) this suggests that life was not merely in existence in those ancient times, but that it was flourishing. Said Schidlowski, 'The ancient earth could have been in (such) a state of global biotic saturation when prolific microbial ecosystems monopolized the totality of suitable habitats.' He went on to note in support of this contention that in recent years numerous discoveries have been made by paleontologists of 'impressive fossil (bacterial) carpets' (known as stromatolites) from rock formations exceeding 2.5 billion years in age.

Schidlowski may or may not have over-interpreted his data. But regardless of this, the information brought back with the carbon time machine is just one more piece of the burgeoning evidence of the proliferation of our ancestors' bacterial life on earth almost 4 billion years ago.

CHAPTER 7

CHILDREN OF TIME

The final emergence of humans from the rest of the animal pack began about 6 million years ago. At that time our ancestors and those of the chimpanzees were one and the same. Our common forebears were tiny (0.9–1.2 m tall), hairy chimp-like creatures, living in the trees which line the rift valleys of central Africa. Then, for reasons no one knows, and which may never be known, a small group of them became geographically, and hence genetically, isolated from all the rest. The deserted chimp-ancestral majority stumbled down a genetic highway to the zoo, scarcely increasing in brain size, and remaining for ever quadrupedal. In contrast, our ancestors followed a ragged path, fraught with perils and extinctions, becoming upright two-legged walkers over 4 million years ago, eventually sporting brains three or four times larger than those of our chimpanzee cousins today. While the details of the final steps to humanity are still not known, the general outline has become clear in the past 30 years through a revolutionary combination of the work of paleoanthropologists and geologists on the one hand and a new breed of scientists, geochronologists, on the other hand.

In unravelling the details and the time scale of our descent from our chimp-like ancestors, the *sine qua non* is, of course, the discovery and scientific characterization of the ancestral fossils by the paleoanthropologists. This, however, is only the first step in solving the broken jigsaw puzzle of our evolution. To reassemble the puzzle, we also have to know, somehow, the ages of the pieces (i.e. the fossils), because our jigsaw is a puzzle in time.

So the $64 000 question occurs: how do we date our fossilized human ancestors? At the outset we should note that the well known carbon-14 dating technique plays no role whatsoever in our story. The short half-life of carbon-14 (a little under 6000 years) dooms it to extinction (in practical terms) in about 50 000 years. But essentially modern man had appeared on earth at least 100 000 years ago; so carbon dating has nothing to say about the time scale of our journey from primitive ape to *Homo sapiens*. It did of course revolutionize our knowledge of the timing of the emergence of civilization in the past 10 000 years.

Dating our human ancestors

The fossils of our ancestors (whom we hereafter refer to as hominids) are not themselves directly dated, for two reasons. Firstly the actual fossils are not ideal subjects for modern dating techniques. They turn out to be unreliable radioactive clocks. Secondly, modern governments guard their hominid fossils jealously, usually keeping them locked away in museums. The last thing that they are willing to do is to let geochronologists dissolve or vaporize pieces of these national treasures in their dating measurements. The ages of hominid fossils, in fact, are found by dating volcanic rocks lying just above or below them in the stratigraphic column. Luckily for us there is no shortage, usually, of such volcanics, because much of our evolution occurred in the rift valleys of Africa. These rifts represent fracture zones along which the continent is breaking apart, and for millions of years they have been the loci of explosively erupting volcanoes. As a result, the layers of sedimentary rock in which the fossil hominids are found today are punctuated by thin layers of volcanic ash which are ideal for dating. These ash layers, called tephra by geologists, are therefore like milestones in time, awaiting the geochronologist's eagle eye.

Potassium–argon dating

The technique originally used to date the volcanic ashes associated with hominid fossils is called potassium–argon dating. In principle, it works exactly like the very first radioactive clock which, you will remember, was proposed by Ernest Rutherford in 1904 (see p 32). The young genius pointed out that the radioactivity of uranium generated the gas helium; so, if one measured the amounts of helium and uranium in a mineral, knowing the rate at which a given amount of uranium produces helium, we could easily (in principle) calculate the mineral's age. While this method has never worked well, because of the Houdini-like escape of helium from minerals, there is now a superbly reliable clock, called potassium–argon dating, which faithfully follows Rutherford's recipe unless it is seriously disturbed. This clock, invented by Alfred Nier (whom we met in chapter 3) and his student Tom Aldrich in the late 1940s, and its modern variant argon–argon dating invented in the 1960s by Craig Merrihue and Grenville Turner, have enabled geochronologists in the past 30 years to erect an astonishingly accurate volcanic time scale against which to view human evolution.

The potassium found in many rock-forming minerals contains a tiny amount of the radioactive isotope potassium-40. (You are actually being irradiated by this isotope in your bodily fluids right now.) This radioactivity generates the gas argon-40, at a known rate. Therefore, *à la* Rutherford, we can see that, if we measure the amounts of the potassium-40 and argon-40 isotopes in a mineral, and if we know (as we do from nuclear physics, essentially from the half-life of potassium-40) how quickly a given amount of potassium generates argon, then we can immediately calculate the mineral's age. It's as simple as that, in principle. In practice it's far more difficult. Very sensitive analytical techniques are needed for the precise measurements of the potassium and argon. In some circumstances, argon may be lost or gained by a mineral, which would give too young or too old an age, respectively. But over the years, the techniques of measurement and interpretation have been greatly refined, so that, in the hominid evolution time interval, ages can be measured with errors of only a few per cent.

The first pioneering potassium–argon dating of hominid-associated volcanics was carried out at the University of California at Berkeley in the late 1950s by Jack Evernden and Garniss Curtis using a mass spectrometer, designed by their colleague John Reynolds, to detect the microscopically small amounts of argon in the volcanic samples. Their measurements eventually showed that the renowned fossil, *Zinjanthropus*, which had been discovered at Olduvai Gorge in Tanzania by Mary Leakey, and named by her husband Louis, was about 1.8 million years old. This stunning result for the heavy-browed large-toothed skull resounded around the world. The age itself has withstood the test of time, but *Zinjanthropus* is no longer considered to be on the direct human (i.e. *Homo*) lineage, being now thought to be an evolutionary offshoot (called *Australopithecus boisei*) that became extinct over a million years ago (see figure 7.1).

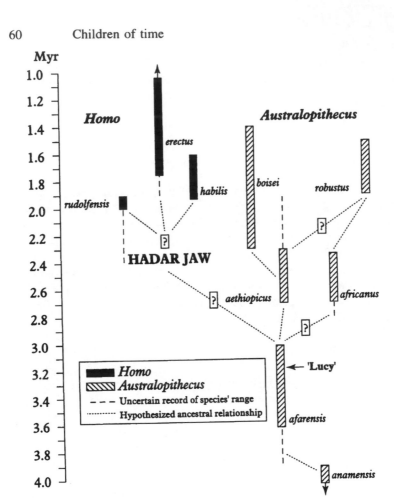

Figure 7.1: *Evolution of human ancestors (after W Kimbel et al).*

Argon–argon dating

In 1965, at the University of California at Berkeley, Craig Merrihue and Grenville Turner (now at the University of Manchester, England) proposed a major improvement in the potassium–argon dating technique. Since potassium is a solid and argon is a gas, these two chemical elements had traditionally been measured in totally different chemical analyses on different samples of the ash. Merrihue and Turner showed that it was possible to measure both elements simultaneously in a single analysis of one sample. They did this very beautifully by converting the potassium in a mineral into an argon isotope known as argon-39, by irradiating the mineral in a nuclear reactor. Then they could melt the mineral to release the argon-39 and the argon-40 which had accumulated over the millions of years in the mineral. And now, in the mass spectrometer, one could simultaneously measure both the argon-40 and the argon-39. But since the argon-39 was produced from the potassium by the reactor, it was in fact a measure of the potassium. So one now knew both the argon-40 and the potassium contents of a sample and so could calculate its age.

Single-crystal laser dating

The final step in refining the potassium–argon method for hominid dating was taken in Toronto by myself, Chris Hall and Yotaro Yanase in the mid-1980s when we introduced continuous lasers into the argon–argon method. A significant problem plaguing the argon–argon dating of hominid-associated volcanic ashes was the frequent occurrence of older contaminants in the ashes. Inevitably, during a violent volcanic eruption, old rocks and minerals are ripped off the walls of the volcano and contaminate the layers of ash. Geochronologists who come along millions of years later and extract crystals of the potassium-rich mineral sanidine for dating cannot be sure how much contaminating older sanidine is in their mineral concentrates. Equally problematical is that they also don't know how old the contaminants are. So, if you need to use a hundred or a thousand sanidine grains for one argon–argon analysis (as was the original practice before our work) your resulting age could be grotesquely wrong—much too old. We therefore showed how it is possible to melt a single tiny (millimetre-sized) sanidine crystal with a high-powered continuous laser and measure the minute amounts of argon-39 and argon-40 released to get the precise age of that crystal. With this approach you can then date, say, 30 such single crystals of sanidine one at a time and see whether you have a single population of ages. If in fact the ages all agree within experimental error, then the average of those 30 ages is the age of the ash. Frequently, however, we find several clusters of ages in an ash which would have completely invalidated an age measured in the old way using many crystals per run. With the laser, however, acting as our discriminator we have no trouble avoiding the contaminants and extracting the correct age for the ash.

The family tree

In the past three decades, then, initially via the potassium–argon method, then with the argon–argon approach and now predominantly with the single-crystal-laser dating technique, a remarkably precise volcanic time scale has been outlined against which spectacular fossil hominid discoveries can be arranged in time. Such an evolutionary sequence is shown in figure 7.1. It should be emphasized that this is a simplified approximation to what will eventually be found to have occurred. Following the traditional approach, a simple set of lines has been drawn joining successively younger fossils, with very little branching. Undoubtedly, as more hominid fossils are found and dated, a more complex human family tree will emerge.

If we scramble chimp-like along the branches of this tree we see it begins (figure 7.1) just over 4 million years ago with a creature called *Australopithecus anamensis* whose fossilized fragments were discovered in the summer of 1995 in northern Kenya by a team led by Meave Leakey, wife of Richard Leakey. A thigh bone showed that *anamensis* was an upright walker. Single-crystal argon–argon laser dating of associated volcanics by Ian McDougall of the Australian National University demonstrated *anamensis'* age fell in the 3.9–4.2 million year range. The previously oldest-known fossil from a hominid was a thigh bone from Ethiopia, dated in the Toronto laboratory at approximately 3.9 million years.

Next we meet the best-known fossil of them all, 'Lucy' (figure 7.2). Discovered by Don Johanson and Tom Gray in Ethiopia near the Awash River, Lucy became famous because of her age and the fact that over 40% of the complete fossil was recovered. About 1 m tall and upright walking, Lucy was assigned the more scientific name *Australopithecus afarensis* by Johanson and his colleague Tim White of Berkeley. Single-crystal argon–argon dating by Bob Walter of the Institute of Human Origins showed that Lucy was a little less than 3.2 million years old. His analyses also showed that the *afarensis* species survived in this region of Ethiopia for at least a million years, from 3.9 million to 2.9 million years ago.

And now, as we leave *afarensis* behind us in the tree at 2.9 million years ago and move a few hundred thousand years towards the present

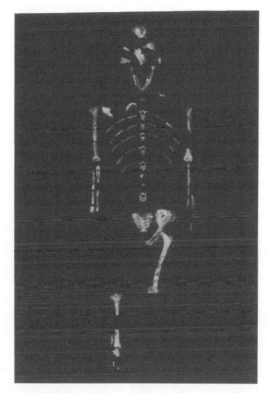

Figure 7.2: *Lucy's fossil, almost 3.2 million years old.*

we realize at 2.3 million years ago that we must have somehow stepped off the *Australopithecus'* branch and are taking a bough towards humanity. For the creatures we are about to meet now (on the so-called *Homo* branch) are distinctly different, becoming larger brained and making stone tools. We can still see the australopithecines on a nearby branch or two, but they're still not making tools and their brains don't seem to grow much larger with time. At about 1.9–1.8 million years ago we meet *Homo rudolfensis* and *Homo habilis*, and, perhaps almost immediately after them, *Homo erectus*. By a million years ago we suddenly realize our cousins, the Australopithecines, have gone, disappeared. They are extinct. So too have *Homo rudolfensis* and *Homo habilis* died out. But *Homo erectus* carries on, gradually evolving in some uncertain way into *Homo sapiens neanderthalensis* and *Homo sapiens sapiens* who appear in the last 200 000 years or so. Finally the neanderthals die out or are absorbed, and *Homo sapiens sapiens* survives alone, an ape produced randomly by natural selection and 4.5 billion years of time, who had no trouble typing the works of Shakespeare.

Crossing the Rubicon

I have just said that by 2.3 million years ago our human ancestors had split off from the *Australopithecus* lineage to form the *Homo* line. How I can say this so confidently about such a crucial step towards humanity makes an interesting example of our search for lost time, and the enormous effort that it entails.

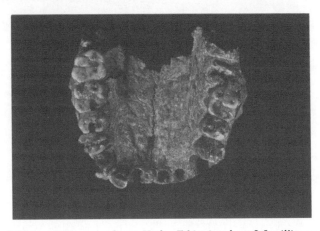

Figure 7.3: *Fossil Homo jawbone, Hadar Ethiopia, about 2.3 million years old.*

On 2 November 1994, a fossilized upper jawbone (figure 7.3) was found in Ethiopia, not far from the Lucy site, by the Berkeley Institute of Human Origins hominid expedition led by Don Johanson, Bill Kimbel and Bob Walter. The fossil was readily identified as coming from the *Homo* lineage, although the precise species of *Homo* (i.e. *habilis*, *rudolfensis* or what?) could not be determined. However, after examining the other types of fossil seen in the associated rocks, the team members were excited because it seemed that the fossil jawbone might represent the oldest *Homo* specimen ever found. Not only that— primitive stone tools were found at the site. The fossil was located stratigraphically less than a metre below a volcanic ash. Samples of the ash therefore were collected for dating by the expedition's chief geologist, Bob Walter, who brought the material to the Toronto laboratory for single-crystal laser argon–argon dating. Because of its chemical composition, the ash contained no sanidine, the mineral of choice (rich in potassium) for the dating technique. As a consequence, the dating was done on a mineral called plagioclase which was extremely low in potassium. Dating such low-potassium minerals was going to be tough. It was as though we had to date potassium-rich sanidines only 10 000 years old, something we would have said was absurd only 3 years earlier because of the minute amount of argon that we had to measure. However, my colleagues Norman Evensen, Pat Smith, Yangshao Chen, Qiang Hu and I had just spent 2 years refining the laser technique so that we could extend the single-crystal approach down to dating sanidines as young as 10 000 years. The system was therefore in just the right shape to date these difficult plagioclase crystals. So 84 single crystals were dated, of which eight were obviously contaminants ranging in age from about 8 million to 27 million years. The remainder gave an age of 2.33 million years with an uncertainty of about 70 000 years. The expectations of the expeditionary team were therefore amply fulfilled, since the fossil was up to 0.5 million years older than any other well dated *Homo* and narrowed the uncertainty in the date of the final divergence of our human (*Homo*) line from the australopithecines to most probably between 2.9 million and 2.3 million years ago (figure 7.1). Since the stone tools were also found at this site, the results also very strongly suggested that creatures of the *Homo* line were making and using such stone tools about 2.3 million years ago.

The oldest tools ever found are about 2.5 million years old and were dated in 1997 by members of the Berkeley Geochronology Centre using laser probe single-crystal dating. Those tools, however, which were found in Ethiopia, are not associated with hominid fossils; so it is not known who made them.

CHAPTER 8

DINOSAURS, METEORITES AND ALL THAT JAZZ

It was, of course, inevitable that the search for lost time would bring us to the dinosaurs, the question of their extinction, and the equally interesting question: was their extinction merely one episode in a series of megadeaths which strike life on earth every 26 million years or so? If so, then what on earth could be responsible for these periodic holocausts? Or perhaps we are just asking the wrong question: it's not what on earth so much as what in the solar system is such a deadly visitor? Is the earth in fact 'attacked' by meteorites every 26 million years?

Ironically, in the light of all the recent interest in meteorite impacts, it is interesting to recall that it is not so long since highly intelligent people were extremely sceptical of the very idea that lumps of stone and/or iron were falling from time to time on the earth from outer space. Among the discoveries that helped to put an end to such scepticism was that made in the 1930s by an eccentric English explorer, who found quite serendipitously, in the sands of Arabia, convincing evidence of the recent fall of an iron meteorite.

Worlds in collision

'I could more easily believe that two Yankee professors would lie than that stones would fall from heaven,' so, according to legend, said US President Thomas Jefferson when told of a reported meteorite fall in the USA. Despite Jefferson's disbelief, many hundreds of meteorites have been found scattered over the earth's surface. Chunks of stone and iron–nickel alloy, the vast majority were traditionally considered to be visitors from the asteroidal belt which lies between the orbits of Mars and Jupiter. It has been surmised that a very few may have come from Mars, and one, recovered not long ago in Antarctica, is probably a piece of the moon. In recent years, as we shall see later in the chapter, it has been suggested that many may have come from the Oort cloud of comets (see p 76).

Meteorites are usually found lying loose on the earth's surface, perhaps turned up by a farmer's plough.   Ancient man, in fact, used iron meteorites as a source of metal.   Despite the widespread occurrence of meteorites, however, remarkably few actual meteorite craters are known.   A meteorite crater is only considered reliably identified if meteorite fragments are found in or near the structure.   Using this criterion, only 11 definite meteorite impact craters are known on Earth (figure 8.1).

Figure 8.1: *Impression of meteorite impact.*

Perhaps the most remarkable (certainly the most romantic) impact crater discovery was made in the 1930s by the eccentric Englishman Harry St John Philby, father of the notorious spy Kim Philby. His announcement of it was made formally on 23 May 1932, to a distinguished audience which had gathered at a meeting of the Royal Geographical Society in London.

St John Philby was an extraordinary individual, one of the last members of that breed of Englishmen, such as Sir Richard Burton and T E Lawrence, who were irresistibly drawn to remote corners of the world. After being captain of his school and winning high honours in modern languages at Cambridge, Philby had served in the civil service in India, before becoming fascinated with Arabia. He had become an authority on that land and the Royal Geographical Society members undoubtedly waited keenly for Philby to begin what was to be his fourth lecture to them over the years.

His talk was entitled 'Rub al Khali—an account of exploration in the Great South Desert of Arabia under the auspices and patronage of His Majesty Abdul Aziz ibn Saud, King of the Hejaz and Nejd and its Dependencies'. Philby delivered an enthralling account of his conquest of the great Rub al Khali desert. With 18 Arab companions and 32 camels, Philby covered 2900 km in 90 days. His purpose was to do general mapping, to be the first man to cross the Rub al Khali from east to west and, most excitingly, to discover two legendary lost cities: Magan and Wabar. At Wabar, he hoped to find a mythical lump of iron 'as big as a camel'.

Philby achieved the desert crossing brilliantly, but his fondest dreams, of discovering ancient kingdoms with riches rivalling Tutankhamen, were shattered. At the reputed site of Magan, all Philby found was a beautifully built, extremely deep well of 'excellent sweet water'. At what was to have been 'the legendary city of Wabar', Philby recounted 'the moment I had longed for for 14 years, and I found myself looking down on the ruins of what appeared to be a volcano! So that was Wabar, the city of a wicked king destroyed by fire from heaven and henceforward inhabited only by semihuman monomembrous monsters... this locality failed to produce the slightest vestige of human occupation, temporary or otherwise, modern or ancient.

'The nearest water was 10 miles [16 km] away at Faraja, and the river I had hoped to find was nowhere to be seen. The secret of Wabar was out at last, and all that remained to do was to search for the iron as big as a camel.' While to Philby such a search for a chunk of iron in the desert seemed like looking for a needle in a haystack, his men did find a piece of the metal. True, it was more like a rabbit in size than a camel, but it excited Philby, because he had decided that the 'volcano' they had stumbled on at Wabar was a meteorite crater and that the iron rabbit was a fragment of the meteorite.

Philby's investigation of the site revealed two large craters 'whose black walls stood up gauntly above the encroaching sand like the battlements and bastions of some great castle. These craters were respectively about 100 yards [90 m] and 50 yards [45 m] in diameter, sunken in the middle, but half-choked with sand, while inside and outside their walls lay what I took to be lava in great circles where it seemed to have flowed out from the fiery furnace'. Three more craters were found nearby.

Philby's speculation was correct. L J Spencer of the British Museum was able to confirm without doubt that the iron rabbit was part of an iron meteorite. As Philby told his audience, 'But Wabar, while it has failed to confirm an Arabian legend, has made amends for such failure by providing the necessary proof of a scientific theory.' The theory, of course, was that meteorite craters do exist on earth.

Philby's account also sheds beautiful light on the difficulties of interpreting legends. The legend of Wabar was half-right; fire had come from heaven, as legend had it. Undoubtedly, wandering Bedouin had seen the spectacular impact and had passed on the story through generations.

By a remarkable piece of serendipity, Harry St John Philby had helped to show that, in a sense, 'worlds' have been in collision. How ironic it is that ideological worlds collided within his son's brain, the encounter probably beginning at Cambridge at about the time of Wabar's discovery. Young Kim Philby graduated from Westminster and Cambridge in his father's footsteps and rose to become head of the anti-Soviet department of Britain's MI6. But all the time his allegiance belonged to the Soviets. While his father before him had become a Muslim, Kim Philby became a Soviet citizen. When his father died in 1960, Kim Philby had carved on his tombstone: 'The greatest explorer of them all'.

Nemesis

Just as the very act of appointment of a professional soccer coach virtually dooms him to dismissal, so are the days of most new biological species numbered from the moment of their emergence via the evolutionary process. Said Charles Darwin, 'The greater number of species in each genus, and all the species in many genera, have left no descendants but have become utterly extinct.'

In a review, published in the *Proceedings of the National Academy of Sciences (USA)*, Professor David Raup and Professor John Sepkoski, Jr, of the University of Chicago, agreed with Darwin. They wrote, 'Virtually all species of animals and plants that have ever lived are now extinct, and the known fossil record documents some 200 000 such extinctions.' However, the Chicago workers disagreed with Darwin in an important point. The English genius had gone on to say, 'The old notion of all the inhabitants of the earth having been swept away by catastrophes at successive periods is very generally given up....' In complete contrast, Raup and Sepkoski claim that their review of the fossil record of evolution shows it to be punctuated by brief episodes of mass (although not total) extinctions, something that many geologists have long since accepted. But, and this is the dramatic part of their paper, they claim that statistical analysis of the evolutionary record shows that these episodes of mega-extinction have occurred, with remarkable regularity, every 26 million years.

Their work is the outcome of great improvements that have been made during the past quarter-century in the dating of rocks and fossils. These have made it possible to put reasonably precise ages to the divisions of the geological time scale that had been worked out by geologists in the last century (figure 8.2). Without a clear knowledge of the ages of the fossils involved, of course, any search for periodicities in the extinction record would be foolish. This would have been the case in Darwin's day, for Darwin had no idea whether the Ordovician creatures (see figure 8.2) for example lived and died 200 million, 300 million or 500 million years ago. (An important question still remains as to whether the geological time scale is known accurately enough even now for a statistical study such as that of Raup and Sepkoski to be acceptable.)

Their study began with Sepkoski making a compilation of the times of existence and extinction of 3500 families of marine vertebrates, invertebrates and protozoan (single-celled) creatures.   This was published in 1982, but when Raup and Sepkoski began to analyse this extinction record for periodicities, they realized that the data were of a very variable quality. It was also clear to them that the geological time scale is less precisely known the further back in time one goes. With these factors in mind, therefore, they restricted their search to simply the last 250 million years or so of history and threw out what they considered to be unreliable data. This resulted in them analysing the extinctions of 567 marine families from late Permian times (see figure 8.2) to 10 million years ago.

By representing their data graphically (figure 8.3) they were easily able to pick out 12 peaks of extinction. Several of these were already well known. The most renowned was that occurring at 65 million years ago which corresponded to the elimination of the mighty dinosaurs from the face of the earth, together with about one quarter of the earth's animal population.

Having identified these times of mass extinction, Raup and Sepkoski then sought a fixed cycle which would most nearly agree with these times. As may be seen in figure 8.3, the agreement was remarkably good with a cycle of 26 million years although the agreement is not perfect. However, on balance, the agreement is intriguing.

The Chicago researchers then performed elaborate computer-based statistical analyses of their data to see how likely it was that such an apparent periodicity could have occurred by chance. Their conclusion— the probability that such a fixed cycle occurs randomly is less than one in a hundred. They said, 'It seems inescapable that the post-late Permian extinction record contains a 26 million year periodicity.' This of course is presuming that the ages assigned to the fossils and the time scale in their work are correct.

Supposing now that the earth's creatures really are subject to a holocaust every 26 million years, Raup and Sepkoski naturally wondered about its cause. Since there are no known biological or geological processes that yield such a cycle, they speculated that the agent would be extraterrestrial.

Millions of years before present	Era	Period	Epoch
0— 2—	Cenozoic	Quaternary	Recent Pleistocene
5— 25— 35— 55—		Tertiary	Pliocene Miocene Oligocene Eocene Paleocene
65—	Mesozoic	Cretaceous	
135—		Jurassic	
205—		Triassic	
250—	Paleozoic	Permian	
280—		Pennsylvanian (Carboniferous) Mississippian	
350—		Devonian	
410— 440—		Silurian	
		Ordovician	
500—		Cambrian	
540—	Precambrian		
3300— 4000— 4600—			

Figure 8.2: *Geological time scale.*

Physical events	Biological events
Ice Age	Modern man Early man
Cascadian mtn building Alpine mtn building Opening of North Atlantic	Hominids Ancestral dogs & cats Ancestral apes Ancestral horses, cattle & elephants First primates
Laramide mtn building (Rocky Mountains) Nevadan mtn building	Extinction of dinosaurs First flowering plants
	First birds Peak of dinosaurs
	First dinosaurs First mammals
Appalachian mtn building	Extinction of many Paleozoic species Rise of reptiles
	Coal-forming forests
Hercynian mtn building	First reptiles
Acadian mtn building Caledonian mtn building	First amphibians First insects First trees First air-breathing animals
	First land plants
Taconic mtn building	First vertebrates (fish) First corals
	Earliest abundant fossils (trilobites & graptolites)
	Scanty fossil record
	Algal & bacteria remains Oldest rocks (Northern Canada) Origin of earth

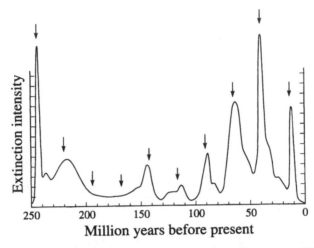

Figure 8.3: *Suggested 26 million year extinction periodicity (after R Muller).*

There was also another good reason for looking skywards. In recent years, a meteorite impact theory of extinction has been championed by Luis Alvarez (Nobel Prize in Physics, 1968), his geologist son Walter and several Berkeley colleagues. These scientists have argued that the spectacular extinction of the dinosaurs and many other creatures about 65 million years ago, at the Cretaceous–Tertiary boundary (figure 8.2), was caused by the devastating collision with the earth of an asteroid about 10 km wide travelling at about 80 000 km h^{-1}.

Now, the idea that mega-extinctions might be produced by meteorite impacts is not new. It was, for example, used a decade before the Berkeley workers by Digby McLaren (former head of the Geological Survey of Canada) to explain a late Devonian (figure 8.2) extinction event. What was new, however, was an ingenious geochemical argument used by the Alvarez group.

On their theory, the colossal impact (equivalent in energy to thousands of millions of Hiroshima atomic bombs) would eject vast amounts of dust into the stratosphere, causing the whole earth to be surrounded by a heavy veil. The sun's rays would be excluded from the earth's surface by this veil for the few months to years required for the dust to settle. In the eerie darkness of this 'long night', the photosynthetic process would be shut off and many plants would die or at least stop growing. Many animals dependent on such plants would die and many carnivorous animals dependent for food on these herbivores would die. A similar type of chain reaction of extinctions could be expected in the sea. (The so-called 'nuclear winter' which has been predicted by Carl Sagan and others to follow a nuclear war is a variant of the Alvarez group's 'long night'.)

What lifted this Alvarez scenario from the realms of pure speculation was their claim to have discovered in the Appenine Mountains, about 160 km north of Rome, the partial remains of the collapsed dust cloud in a thick layer of clay about 1.27 cm thick lying in a column of sedimentary rocks which span the Cretaceous–Tertiary boundary, i.e. the time of the dinosaur extinctions. The rocks below the clay layer contain various kinds of fossils which are not found in the rocks above the clay. Clearly the extinction of numerous species occurred at about the time the clay layer was formed.

The Berkeley contribution was to show that the clay did not simply mark the time of a mega-extinction event, like a bookmark in the pages of natural history. By making measurements of extraordinary sensitivity, they showed that the rare metal iridium was present in the clay to a much a higher level of concentration than usual for terrestrial rocks. But they knew that iridium occurs much more commonly in meteorites; so they concluded that here was the very stuff of the meteorite itself. Yet this extraterrestrial material occurred in the clay exactly at this location of a major extinction event on earth. Therefore this was either a fluke or, as they concluded, here was the geochemical fossil record of the fatal impact—a beautiful idea.

As is usual in science, the new theory was liked by some but fiercely opposed by others. But rightly or wrongly attributed to the fall of a giant meteorite, there is no doubt that the 'iridium anomaly' itself exists in sediments at the Cretaceous–Tertiary boundary in many locations on earth. In a later review the Berkeley workers noted that it 'has been found in 50 sections worldwide'.

If it is possible that one major extinction event (the Cretaceous–Tertiary event) was caused by meteorite impact then it is natural to wonder if all ten of the extinctions in the fixed cycle suggested by Raup and

Sepkoski were due to such a cause. At first sight, this seems highly unlikely. Most meteorites are thought to be solar system rubble perhaps ejected by collisions from the asteroidal belt which lies between Mars and Jupiter. Such a process would not be expected to blast the earth with giant meteorites every 26 million years.

However, two groups of astronomers have recently come up with an ingenious method of pelting the earth at regular intervals. They have suggested that the sun may have a 'dark companion', a small dark star moving in an elliptical orbit around it. The star could be dark either because it was not quite large enough for thermonuclear fires to begin in its interior, or because it was already burnt out. In either case, they note that if such a star reached a maximum distance from the sun of about 2 light years, its orbit would bring it hurtling into the inner solar system every 26 million years. In doing this, it would pass through the so-called Oort cloud of comets which is supposed to surround the known solar system. The intruding star's gravitational field would unleash some of these comets from their normal paths hurling perhaps 20–200 of them towards the earth in a period of about a million years. It is suggested that such periodic bombardment showers could be the cause of the extinction periodicity of 26 million years.

The name Nemesis has been proposed for this marauding star by one group of astronomers, Marc Davis and Richard Muller of Berkeley, and Piet Hut of Princeton, after the Greek goddess who 'relentlessly persecutes the excessively rich, proud and powerful'.

Whether the iridium anomalies and the extinction episodes are in fact all produced by giant meteorite impacts is not yet clear. Whether the extinction traumas are cyclic with a period of 26 million years is also not proven. Whether Nemesis exists and is plunging towards a new attack on earth in about 10 million years time (its calculated next time of approach) remains to be seen. But if this scenario is correct, then our evolutionary descent from primitive structures to reasoning human beings was one long roller-coaster ride through a cosmic shooting gallery.

The evidence and arguments put forward by the Alvarez team convinced many people that at least the 'dinosaur extinction' at the Cretaceous–Tertiary boundary was in fact triggered by a giant meteorite impact.

Others, however, said, 'Where's the 150–200 km crater that should have been created by the impact 65 million years ago?' For a number of years this did seem to be a good question, until 1991 when Alan Hildebrand (now at the Geological Survey of Canada, Ottawa) and a group of co-workers presented detailed evidence of the existence of just such a crater on the Yucatan Peninsula of Mexico. The crater of 180 km diameter lies hidden under about 1 km of limestone deposited in the 65 million years which have elapsed since impact. (Ironically, the existence of this buried crater had been proposed 10 years earlier by two oil geologists, Glen Penfield and Antonio Camargo, who noticed curious circular features in maps of the earth's gravity and magnetic fields in the area. Their ingenious interpretation, however, had been ignored by meteorite students for a decade.) In 1992, glass recovered from the crater by deep drilling was dated at about 65 million years old by the laser probe argon–argon method (see p 61) by Carl Swisher and his colleagues at the Berkeley Geochronology Center. If the glass was formed from rock melted by the impact, this age confirms the existence of a presumed meteorite crater of precisely the right age and just the right size to fit the Alvarez impact scenario, a beautiful outcome which has swung even more scientists to this viewpoint.

CHAPTER 9

ATOMIC 'REACTOR' OPERATED 2 BILLION YEARS AGO

One of the more extraordinary consequences of the enormous age of the earth (4.5 billion years) and the relatively short half-life of uranium-235 (about 700 million years) is that at least one natural nuclear reactor sprang into spontaneous operation about 2 billion years ago. The brilliance of a French researcher, in search of lost uranium, took modern geochemists into the steaming heat of equatorial Africa in southwest Gabon in search of lost half-lives.

Imagine a brilliant cryptographer poring for years over an ancient piece of stone inscribed in some unknown language. Imagine his delight turning to amazement when he cracks the code and hesitatingly transcribes the following words: recipe for a home-reactor—take one bathtub full of uranium ore; add water and stir. Then his puzzlement turns to a smile; he does a few brief calculations and he has the remarkable answer. The creature that carved the stone lived more than 600 million years ago.

Of course, this is a fanciful tale. There was no thinking being alive on our planet 600 million years ago. However, an extraordinary discovery about an ancient nuclear reactor did really take place in the 1970s. And the reactor was operating not 600 million years ago, but about 2 billion years ago. The 'cryptographer' was the French scientist H Bouzigues. He made the key initial observation on the modern-day Rosetta Stone in 1972 in a nuclear fuel-processing plant at Pierrelatte in southern France, about 160 km north of Marseille.

Bouzigues' stone was a piece of uranium ore. Some of the uranium had been extracted from this and converted into the gas uranium hexafluoride, sometimes known as hex. Bouzigues did a routine mass spectrometer analysis on the hex and found a slight oddity in his result. He knew that natural uranium is almost but not quite 100% uranium-238. Less than 1% is the critical uranium-235 which is needed for fission reactors. What the scientist had found, in fact, was that there seemed to be a tiny amount of uranium-235 missing from his hex. The missing amount was, in fact, so small that many a scientist would have ignored it, i.e. put it down to slightly more than usual experimental error because less than 0.5% of the uranium-235 atoms was missing. But Bouzigues, like Fleming with his mouldy Petri dish, decided to pursue this small peculiarity.

The history of the enigmatic hex was traced back through the whole of its processing at Pierrelatte with the conclusion that the uranium as it arrived in France was already anomalous. No process in the plant had caused the minuscule uranium-235 loss. Events shifted rapidly to Gabon, a country on the equator in west Africa, because the curious ore had come from an open-pit uranium mine at Oklo about 80 km northwest of Franceville in southwest Gabon.

Almost overnight, the uranium mine was transformed into a scientific gold mine: widespread sampling of different pockets of ore revealed that some localities were enormously deficient in uranium-235. In some cases, close to half the isotope was missing. Hardly the kind of uranium fuel for a Candu reactor!

The big question, of course, was: where had all the uranium-235 gone? It could not simply have been dissolved and washed away or removed by any simple natural chemical process, because the uranium-238 isotope was mostly still in place. Only the uranium-235 had been dramatically depleted and such a separation of uranium isotopes is very difficult to produce chemically (luckily for us; otherwise atomic bombs would be easy to make).

The answer to the puzzle came very quickly following a battery of chemical and physical tests on the ore body; in at least six different sites, the ore body had been operating as a spontaneous self-regulating nuclear reactor.

When the fission of the uranium-235 takes place in a modern reactor, a variety of so-called fission products results. The chemical type and isotopic composition of these point to neutron-induced fission. And it was the study of such isotopic fingerprints in the minerals comprising the Oklo deposit that proved beyond any doubt that a chain reaction of uranium-235 fissions had been operating successfully long before any advanced life forms had evolved on earth.

Did the deposit flare up briefly into action or did the reactor run for years? It is hard to pin down the operating lifetime precisely, but it is clear the reactor ran, perhaps intermittently, for thousands of years, perhaps several hundred thousand. It probably ran at a power level of 10–100 kW, enough to provide the electric power for a modern home at typical rates of conversion of heat to electricity. The total estimated amount of missing uranium-235 indicated that the total energy produced by the Oklo reactor in its lifetime was about equal to that delivered in a year by about six modern power reactors.

The obvious question to ask now is: why aren't little reactors running spontaneously today in the world's uranium deposits? And the simple answer is that there is not so much uranium-235 around now as there was 2 billion years ago, when the Oklo reactor was running. This is because uranium-235 is radioactive and is dying away steadily in amount. So the Oklo reactor could go critical because, 2 billion years ago, there was about seven times as much uranium-235 in natural uranium as there is now.

As the hypothetical cryptographer realized, for about the last 600 million years, the earth's uranium-235 supply has been so low that no spontaneous reactors could spring into action, i.e. no bathtub models would work. Today, the uranium-235 content of natural uranium is so low that it took the combined ingenuities of some of the world's best physicists, chemists and engineers more than two years to produce a functioning reactor in a squash court under the west stands of the University of Chicago's Stagg Field in 1942.

Soon after the extinct Oklo reactor had been discovered, it was realized by several people (including G A Cowan of the Los Alamos Laboratory in New Mexico, who has graphically described the Oklo phenomenon) that the site could potentially provide important information on the question of long-term disposal of the nuclear wastes from modern nuclear power plants. While it should be possible to store radioactive wastes safely for hundreds of years, during which time most of the dangerous radioactivities will have become negligible, there is some question about the secure disposal of the plutonium-239 (a byproduct of the reaction) with its half-life of 24 000 years. How effectively will the proposed deep burial sites retain this material?

Here, the Oklo reactor site provided some interesting information. Detailed microprobe studies of individual mineral grains showed that the plutonium-239 produced in the natural reactor stayed in place and did not escape by some leaching process, even though water must almost certainly have been present in the ore body to act as a moderator of the neutrons. This is a very important observation, and yet one that could never have been predicted by Bouzigues when he was trying to explain his odd mass spectrometer results. However, it should be remembered that this is only one illustration of the behaviour of plutonium over very long periods. It may be that the Oklo environment was particularly favourable for plutonium immobilization.

The question naturally arises: were any natural reactors running in other parts of the earth in the pre-Cambrian era? The numerous uranium deposits in northern Saskatchewan, in Canada, for instance, are certainly old enough, probably ranging in age from 1 billion to almost 2 billion years. While samples from some of these have been analysed, no ancient Canadian reactor has been unearthed. But, almost certainly, a far more detailed study could be carried out on them and on new ones as they are found. Should extinct Canadian or other reactors be found in this way, then it would be very interesting and important, as Cowan has emphasized, to carry out microprobe analyses to see how the plutonium behaved in the possibly different environments.

CHAPTER 10

GULLIVER'S TRAVELS AND MARTIAN MOONS—TIME FOR KEPLER

We shall see in our discussions of Einstein's relativistic theories that space and time came to be interwoven into a single fabric called space–time. But, even in Newton's theories, space and time were already intimately linked. In particular, since Kepler's era it had been known that the *periods* of planets in their orbits bore a curious fixed relation to their *distances* from the sun. In this chapter we shall see how time's entanglement with space fatally punctures an argument advanced by the fantasist Imanuel Velikovsky in his widely read *Worlds in Collision*.

When Jonathan Swift wrote his celebrated *Gulliver's Travels*, published in 1726, he included the following remarkable passage: 'Certain astrologers... have likewise discovered two lesser stars, or satellites, which revolve about Mars, whereof the innermost is distant from the centre of the primary planet exactly three of its diameters, and the outermost five; the former revolves in the space of ten hours, and the latter in twenty-one and a half; so that the squares of their periodical times are very near in the same proportion with the cubes of their distance from the centre of Mars; which evidently shows them to be governed by the same law of gravitation that influences the other heavenly bodies.' What makes this passage remarkable is that in 1726 no moon had been observed circling Mars, and yet in 1877 Asaph Hall with the Washington telescope discovered precisely two moons orbiting the red planet. Did the ironist Swift really know of the existence of the moons of Mars 151 years before they were 'officially' discovered? It was most striking that Swift got right the number of moons. It is also true that these satellites go around their parent in small orbits at relatively high speed, although, as we shall shortly see in more detail, Swift's numerical values were far from correct. Many people have been intrigued by this passage in *Gulliver's Travels*, including Sir George Darwin, the well known theoretical geophysicist and son of Charles Darwin, co-formulator of the theory of biological evolution. Darwin dedicated a small section of his popular book on the tides to Swift's fantasy. Subsequently, Velikovsky saw that the strange affair might provide support for his concept of worlds being recently in collision. For if Mars did, in fact, closely approach the earth in about 700 BC, as Velikovsky claimed, then perhaps its moons were clearly visible and the radii of their orbits, and their periods, might have been estimated. Perhaps someone wrote down these facts in some record now lost. Velikovsky said that there is a 50% chance that Swift just made a lucky guess about the Martian moons. On the other hand, he said, perhaps Swift read of the moons 'in some text not known to us or to his contemporaries. The fact is that Homer knew about the "two steeds of Mars" that drew his chariot; Virgil also wrote about them.'

I would like to suggest (as have others for different reasons) that the source of Swift's 'information' was not some long-lost record from 700 BC. Swift's idea came from Kepler, the extraordinary genius of astronomy, who was conceived on 16 May 1571 at 4:37 am, and born on the 27 December at 2:30 pm. Kepler was a mystic, fascinated by numbers and relations among them, and it was typical of him that he would have sought to record the moment of his conception from his parents. This brilliant man grew up before the modern scientific age, in fact before the term 'scientist' had been coined. His life was surrounded by witchcraft and astrology. His own mother was tried as a witch and Kepler himself acted as lawyer in her defence. One of his jobs was preparing horoscopes for the Duke of Wallenstein. Kepler was as much the astrologer as the scientist. But extraordinary scientist he was, and after many years of undaunted struggle, principally with Tycho Brahe's observations of Mars, he came up with what have long been known as Kepler's laws. We shall meet these again soon, but for the moment we shall see how Kepler came to write of the moons of Mars over 200 years before their discovery, and long even before the publication of *Gulliver's Travels*. By a strange quirk the story involves the great contemporary of Kepler, Galileo. The latter had been announcing a series of spectacular astronomical observational discoveries which he had made with one of the earliest telescopes. However, because of the intense competition which existed between scientists, then just as now, Galileo was in the common habit of announcing his new results in Latin anagrams. That is, the statement of a new result or law would be written in Latin and then the letter order would be scrambled. Thus, if some unscrupulous competitor wished to claim he already knew of your newly announced results, he first of all had to unscramble your anagram. So, when Galileo discovered with his telescope what he thought were two moons of Saturn (but what were in actuality the rings of Saturn) he wrote, '*Altissimum planetam tergenimum observavi.*' This may be translated as 'I have observed the highest planet (Saturn) in triplet form'. Galileo proceded to form an anagram from the Latin words and gave the following sequence of letters to the Tuscan ambassador in Prague: 'SMAISMRMILMEPOETALEUMIBUNENUGTTAURIAS'. When Kepler, who had no telescope at that time, was shown the message he set to work to decode it with his usual flair and came up with the following piece of 'barbaric Latin verse': '*Salve umbistineum geminatum Martia proles*'. Arthur Koestler in his brilliant book *The Sleepwalkers* (not thinking of Velikovsky's work) translates this as, 'Hail burning twin, offspring of Mars.' Said Koestler, 'He (Kepler) accordingly believed that Galileo had discovered moons around Mars, too.' Here then is the origin of the prophetic passage in *Gulliver's Travels*. The 'astrologers' of Jonathan Swift were a reference to Kepler the astrologer and the 'two lesser stars, or satellites' were Kepler's twin offspring. Swift's inspiration, then, was not some unknown ancient manuscript. It was, ironically, a mistaken piece of ingenuity by one of the greatest astronomers of history that triggered Swift's enigmatic passage. No one would have enjoyed the joke more than the great ironist when Asaph Hall announced that in fact Mars does have two moons.

Now it is reasonable to ask: did this man of literature, Swift, know of this astrologer and scientist Kepler? The answer is a resounding, 'Yes'. For his remark, '...so that the squares of their periodical times are very near in the same proportion with the cubes of their distance from the centre of Mars', is a specific reference to Kepler's third law which states that, for planets going around the sun in approximately circular orbits, the square of the orbital period (i.e. $T \times T$) divided by the cube of the average distance of the planet from the sun ($L \times L \times L$) is the same number for all planets. Quickly consider the earth and Mars as examples. The earth's orbital period is 365.25 days while its average distance from the sun is 1.0 astronomical unit. Therefore we have $(T \times T) / (L \times L \times L) = (365.25 \times 365.25)/(1 \times 1 \times 1) = 133\,407.56$ for the earth. On the other hand, the Martian year is 686.98 days while, on average, its distance from the sun is 1.5237 astronomical units. Hence, we find for Mars that $(T \times T)/(L \times L \times L) = (686.98 \times 686.98)/(1.5237 \times 1.5237 \times 1.5237) = 133\,410.43$, which differs from the quantity just found for the earth by two thousandths of a per cent. Therefore, when Swift adopted Kepler's misbegotten offspring of Mars and guessed periods and radii for their orbits, he made the numbers obey Kepler's law (which applies to small moons circling a planet as well as it does to planets circling a sun)—hence the above-quoted remark. For if Swift's satellites were to 'be governed by the same laws of gravitation, that influences the other heavenly bodies' the quantity $(T \times T)/(L \times L \times L)$ for each moon should be the same. We can easily verify this. Swift's inner moon, he said, had $T = 10$ h and $L = 3$ diameters of Mars. These give $(T \times T)/(L \times L \times L) = (10 \times 10)/(3 \times 3 \times 3) = 3.704$. The outer satellite was supposed to have $T = 21.5$ h and $L = 5$ diameters of Mars, so in this case $(T \times T)/(L \times L \times L) = (21.5 \times 21.5)/(5 \times 5 \times 5) = 3.698$, which differs from the value just found for the inner moon by about one sixth of a per cent. Swift, therefore, apparently did choose the numerical features of his moons to satisfy Kepler's law. We may deduce then that Swift had certainly heard of some of this astrologer's writings!

Now it might be argued that the fact that Swift's moons obey Kepler's third law is not because Swift made them do so but because he found these values in some ancient eyewitness record of the moons. They obeyed Kepler's law because they really existed and any existing moons would automatically satisfy the Kepler relationship because that's the way moons are. This argument would be perfectly valid but for one problem, namely, the density of Mars. To satisfy Kepler's law, Swift not only had to choose an L and a T value for each satellite so that they had a common value for $(T \times T)/(L \times L \times L)$ but also had to choose just what this common value was. But to settle on the correct value he needed to know the ratio of the density of Mars to that of the earth. And here is the crunch—in 1726 neither the density of Mars nor that of the earth was known! So Swift must have guessed this ratio and, as a direct result, obtained the value for $(T \times T)/(L \times L \times L)$ of about 3.7 as we saw above. But this means that he guessed a value for the ratio of the terrestrial density to the Martian density which was *six* times too small, i.e. Swift's guess for the relative density of Mars was *six* times too high.

Thus, because Swift didn't know the relative densities of the earth and Mars his imaginary moons only partially satisfy Kepler's third law. Their values of $(T \times T)/(L \times L \times L)$ were equal, all right, but equal to the wrong thing! But, of course, no real moons can misbehave like this and therefore Swift's source was not some ancient observer who saw the Martian moons during a close encounter between Mars and the earth. Swift's source was his imagination.

Asaph Hall himself, discoverer of the Martian moons Phobos and Deimos, 'the dogs of war', suspected for another reason that Swift's guess came via Kepler. For in his paper (1878), 'Observations and orbits of the satellites of Mars', Hall quoted the following reference by Kepler to the possible existence of two Martian moons: 'I am so far from disbelieving the existence of the four circumjovial planets that I long for a telescope, to anticipate you, if possible, in discovering two round Mars, as the proportions seems to require, six or eight around Saturn, and perhaps one each round Mercury and Venus.' Kepler's guesses, which were correct for Mars but not for Saturn, Mercury or Venus, were based on his mystical belief in numbers and the importance of certain shapes of solids. It is interesting that Velikovsky did not give his readers Asaph Hall's much more probable and far simpler solution to the Gulliver enigma, even though it had been publicly discussed by Hall and Darwin.

Curiously enough, Swift, besides 'anticipating' Hall, also foreshadowed Velikovsky's theory of cometary collision, in *Gulliver's Travels*. For Gulliver at one point says, 'Their apprehensions arise from several changes they dread in the celestial bodies. . . that the earth very narrowly escaped a brush from the tail of the last comet, which would have infallibly reduced it to ashes; and that the next, which they have calculated for one-and-thirty years hence, will probably destroy us. For if, in its perihelion, it should approach within a certain degree of the sun (as by their calculations they have reason to dread), it will receive a degree of heat ten thousand times more intense than that of red hot glowing iron; and, in its absence from the sun, carry a blazing tail ten hundred thousand and fourteen miles [about 1.6×10^6 km] long, through which, if the earth should pass at the distance of one hundred thousand miles [about 1.6×10^5 km] from the nucleus, or main body of the comet, it must, in its passage, be set on fire, and reduced to ashes.'

CHAPTER 11

CHAOS AND TIME

In the past decade chaos has been one of the hottest topics in science. Laser physicists, meteorologists, zoologists and physiologists all study it. Could 'chaos' explain 'the arrow of time'? Could it explain the oddity that we all immediately recognize when we see a motion picture run backwards? Is 'chaos' the reason why 'all the king's horses and all the king's men, couldn't put Humpty together again'?

Peter Coveney and Roger Highfield evidently think so and try to explain why in a recent popular book *The Arrow of Time*. While I don't accept their explanation, it is true that chaos and time are intimately related. We shall see that chaos needs time, in the case of the solar system hundreds of millions of years of time, to express itself, i.e. to emerge from the Platonic shadows and take its place as a recognizable actor on the universe's stage, its role literally increasing (at least for a time) exponentially in importance. But we're getting a little ahead of ourselves here and will have to stop and ask ourselves what physicists mean by 'chaos'. Because in physics 'chaos' has a specific well defined meaning which, while obviously related to our usual every-day understanding of the word 'chaos', is much more constrained.

The key concept in chaos is 'sensitivity to initial conditions'. The most beautiful illustration of this term was given by the renowned English novelist and playwright J B Priestley in 1932 in his first play *Dangerous Corner*. When his maiden play was first staged in London, it was a disastrous flop and closed after only five performances. But Priestley was by then already wealthy from the success of his novels (such as *The Good Companions*), so he personally financed a continuation of the production, which took off.

Dangerous Corner is still being staged around the world 60 years later,

Figure 11.1: *1988 playbill for Dangerous Corner (courtesy Shaw Festival).*

long after Priestley's death in 1984. Several years ago, a production of *Dangerous Corner* was staged in Niagara-on-the-Lake, by the Shaw Festival in Canada; writers such as Priestley and Agatha Christie are used from time to time to add some variety to George Bernard Shaw's output. The Niagara peninsula is the southernmost part of Canada jutting into the USA well below the forty-ninth parallel, and its mild climate makes it the wine centre of Canada. This, plus the nearby incredible Niagara Falls, makes the flower-filled restored old town a perfect site for summer theatre.

The 1988 production of *Dangerous Corner* (figure 11.1) at the old St George Theatre began innocently enough in a replica of a fine

English country home. The cast was finely dressed for cocktails, with the men in dinner jackets, and the women in elegant gowns. After some conversation, one of the men went to switch on a stately looking 'wireless' set (radio), the intention being to tune in to a dance band and enjoy an evening of foxtrotting. The wireless, however, remained silent, perhaps because of a broken switch; so the gathering had to forget about dancing and sat around talking.

The originally light mood rapidly turned dark, accusations were shouted and someone died by the end of the play. The curtain fell, the audience applauded long and loud, so the curtain rose again for the players to take their bows. Or so everyone anticipated. But in fact, Priestley had a stunning surprise prepared. When the curtain rose it was not on a sequence of smiling bowing actors; it rose to reveal the very opening scene of the play, exact in all its detail. So the mystified audience took up its seats again and watched in growing bewilderment for several minutes as they saw and heard a perfect replica of the opening of *Dangerous Corner*.

And now Priestley sprang his extraordinary surprise; the same man as before retraced his identical steps to the wireless and turned it on. Only this time the radio gave forth the silky tones reminiscent of Victor Sylvester's dance band, and the various men on stage invited the ladies to dance a foxtrot. After a few steps and some happy laughter, the curtain descended for the last time; the play was now over.

I was so startled by this experience that it was not until I was half-way back to Toronto, driving on the Queen Elizabeth Way, that I realized that Priestley had discovered chaos, about 30 years after its neglected discovery by the French mathematical genius Henri Poincaré, and about 30 years before its rediscovery in 1963 by Massachusetts Institute of Technology meteorologist Edward Lorenz (see figure 11.2)!

For Priestley was illustrating in a wonderful and entertaining way for non-scientists (but also for scientists) the key concept of chaos, namely the sensitivity of a system to initial conditions. Everything was identical (or as nearly as the director could arrange) between the two versions of the opening scene, except for the condition of the on–off switch of the wireless. The first time around; the switch was faulty, maybe the contacts were dirty or maybe a connection was loose. The second

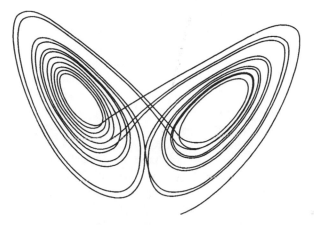

Figure 11.2: *Abstract representation of chaotic weather.*

time around, the switch worked! This minute difference, however, in the behaviour of a tiny component led to two extraordinarily different futures—one filled with hatred and tragedy and one filled with laughter and dance!

This incredibly dramatic presentation by Priestley stunned me. The following year my wife and I returned to Niagara-on-the-Lake to see his *Inspector Calls*. Sure enough elements of chaos feature in it. It was obvious that Priestley was obsessed with time and chaos. I discussed this revelation of Priestley with a number of people, but it was some time before one of them (Norman Evensen) told me that he had seen a book in Edwards' Book Store in Toronto by J B Priestley entitled *Man and Time*. It was 'a personal essay (published in 1964) exploring the eternal riddle: the theories, the philosophy, the scientific discoveries and the everyday'. Here was another Yorkshireman obsessed with time! His beautifully illustrated book runs to 319 pages.

Edward Lorenz's rediscovery of Poincaré's concept of chaos 30 years after the publication of *Dangerous Corner* gives one familiar with the play a peculiar *déjà vu* feeling—something the time-haunted Priestley would have richly enjoyed. Using an early computer, Lorenz was trying to model mathematically the behaviour of the weather (figure 11.2). After building his mathematical equations of the model into the computer, Lorenz would experiment by feeding an initial state of his model to the computer which would then march off calculating the future state of the atmosphere. By starting off from a variety of significantly different initial states, he found not surprisingly that widely differing futures would ensue. Then, on one fateful occasion, Lorenz inadvertently did a Priestley! He took the printed output from the middle of one of these computer runs and used it as the starting point of a brand new run, fully anticipating that the computer would retrace the exact future that it had earlier traced when it went through that atmospheric state. How could it do anything else? Lorenz thought he was doing with his computer the exact equivalent of a human being who, finding a winding trail of footprints on a beach, steps into the midst of a trail (i.e. the early computer output) and places his feet successively into the footprints and follows them to their end point, despite all the twists and twirls around the rocks in the sand. Lorenz was doing a Priestley, only restarting the play in the middle rather than at the beginning.

In actuality, Lorenz set the computer off on its trail of silicon footprints and left for several hours. On his return he was stunned to find that, while his computer had faithfully followed the earlier footsteps for a while, it had eventually begun to diverge in its output from the earlier print-out. What was more, the more its predicted 'weather' diverged from the original, the more it diverged. That is, the two computer trails diverged exponentially (the meaning of exponential will shortly be made clear when we talk about the game of billiards).

Instead of concluding that something had gone temporarily wrong with his computer and forgetting about it, as he might well have done, Lorenz realized that he had discovered something remarkable. For, while his computer was carrying out its calculations using six-figure accuracy, it only printed out its results to three figures. So, even though Lorenz had re-submitted to the computer an exact copy of the earlier output at the selected point as his new starting point, he wasn't injecting quite the right number to the six-figure accuracy that was in the computer's memory in the original run. For instance, a number that the computer was using in the original run in its calculations equal to 137.452 would have appeared in the print-out (and therefore been used by Lorenz as his new starting point) as 137. Lorenz realized that his two calculated futures differed because, if you like, their presents differed very slightly (by the difference between 137.452 and 137) i.e.

the calculated futures of his model atmosphere were extraordinarily sensitive to initial conditions.

In fact, Lorenz found that the difference between these two futures increased fourfold roughly every week, so that after a month the difference had expanded two-hundredfold to three-hundredfold! If his model was remotely like the real weather, there was no hope of accurately forecasting the weather 2 or 3 weeks in the future. The atmosphere would be so sensitive in its response to the slightest variation in its initial temperature state, or its pressure variations over the earth or in the variation in humidity from point to point that accurate long-range forecasting would be a hopeless dream. And that so far as we know is the true state of affairs. Lorenz evocatively called it 'the butterfly effect'. The beating of a butterfly's wings in the Caribbean could trigger a tornado in the midwestern USA weeks later!

Lorenz's publication of his seminal work evoked little response at large and only in the next decade, the 1970s, did other scientists in other fields uncover the profound significance of Lorenz's work for many branches of science. The staggering realization dawned that many systems in nature were described by equations which had this same enormous 'sensitivity to initial conditions' that Poincaré, Priestley and Lorenz had each in his own way so eloquently written about.

This was an extraordinary development conceptually. As we shall see in chapter 12, the discovery of quantum mechanics had shown that in the microscopic world of electrons and nuclei the universe is inherently unpredictable. A still not understood fog of probability had descended on microphysics. Yet it had long been widely believed that in the macroscopic world of our everyday lives determinism reigned. The motions of tennis balls, rockets and planets were phenomenally accurately predictable by Newton's laws. Even Einstein's relativistic modifications of these were deterministic. This philosophy was perhaps best expressed by Pierre Laplace, who was moved to imagine a super-intelligence which, knowing the position and velocity of every particle in the universe, could with perfect accuracy predict the future and reconstruct the past: 'to it, nothing would be uncertain, both future and past would be present before its eyes'.

And while it was obvious that we could never know the present with such perfect accuracy, it was generally believed that in the Newtonian macroworld, if you knew the present state of affairs fairly precisely in a system, then, at least with a super-computer, you could predict the future reasonably accurately and your forecast would stay reasonably close to reality over long periods, perhaps indefinitely, if you even worried that far ahead.

But the great pioneers of chaos had dropped a bomb on this scenario. They had shown that much of the macroworld around us is chaotic. Its future would be exactly predictable if only we knew the present in perfect and total detail. But since we can never attain such perfection of knowledge, then we can never accurately predict these futures. Initially our forecasts will seem faithful, but more and more radically we shall be wrong. Some systems will be more sensitive than others. Some, like the solar system, behave predictably for hundreds of millions of years before possibly wandering off their predicted courses. Others, like the weather, get lost in weeks. But, always given time, an unpredictable future will manifest itself! The image that we get is quite the opposite of Coveney's and Highfield's. Instead of chaos determining the flow of time, time gives birth to chaos!

Curiously, even though the physicist often says that 'chaos comes from non-linear differential equations', which sounds formidably complex, we can in fact get a very simple understanding of how time 'produces' chaos and what we really mean when we say that 'futures diverge exponentially'. Surprisingly enough, the explanation hinges on the game of billiards (or pool or snooker). The rationale was presented by French mathematician Emil Borel, the Soviet mathematician G Sinai and English theoretical physicist Sir Michael Berry.

An outrageous statement

We introduce the ideas with the following outrageous statement essentially due to Michael Berry: '. . . if we neglect the effect of the gravitational field of an electron at the edge of the galaxy on the motion of the cue ball, and if there is no friction to slow down the balls, so that the various balls never stop rolling and colliding with each other and the cushions, then our calculations of the positions and motions of the balls after about two minutes will be hopelessly wrong, even if we are using a super-super-computer to calculate the motions!'

If this could possibly be true, what a beautiful example of chaos it would be. For the presence or the absence of the incredibly small, phenomenally distant electron causes totally different alternative futures. To follow the argument requires no more than elementary arithmetic. We need only know that, when a decimal number, such as 0.1 or 0.01 or 0.001 is multiplied by 10, all that happens is that the decimal point moves one position to the right. Everyone knows this: $0.1 \times 10 = 1.0$; $0.01 \times 10 = 0.1$; $0.001 \times 10 = 0.01$. Armed with this universal knowledge from our pre-teen schooling, we can gain profound insights into chaos and exponentially diverging futures. Remarkably, we shall even get a feeling for what 'non-linearity' means, in a metaphorical way!

When we strike the cue ball it rolls off in a straight line. It's easy to calculate the slight deflection from this straight line that we would get if we introduced an electron at the edge of our galaxy. It is very roughly 0.000 000 000 000 000 000 000 000 000 000 000 000 000 000 000 000 000 000 000 000 000 000 000 000 000 000 000 000 000 000 000 000 1 of a degree! The precise value is not important. Yes. That is, less than a trillion trillion trillion trillion trillion trillion trillion trillionth of a degree! Which is a far, far smaller deflection than we shall ever be able to measure, however long our civilization lasts and however sophisticated we become technically! Here's another way of looking at this unimaginably small error that we would be making in the initial path of the cue ball if we neglected this gravitational tug of the distant electron. If the billiard ball were to travel a distance equal to the diameter of the universe in a straight line, then the error in direction that we are discussing would cause a difference in arrival points at the edge of the universe differing by far, far less than the diameter of a hydrogen atom!

How then could the neglect of such a minuscule error in direction lead to any conceivable significant unpredictability in the motion of our billiard balls, even though we allow the balls to keep rolling for as many collisions as we like, since we are ignoring friction? The answer is where 'non-linearity' comes in. And it enters in the simplest of ways. It arises through the curvature of the surface of a billiard ball (i.e. the ivory surface is curved; its equator is a circle and not a straight line— literally non-linear). The chaotic unpredictability comes in through the

remarkable effect of this curvature during the collisions of the balls.
In total contrast, the side cushions of the billiard table are perfectly
flat. The boundaries of the table are perfect straight lines—linear! As
a result, if we had merely one ball on the frictionless table and set it
rolling, its subsequent motion would not be chaotically unpredictable,
however many times it bounced off the sides. This is because, in
collisions of the ball with the straight cushions, there is *no amplification*
of any initial error in direction. If we approach the side with a 1° error
in direction, we rebound with a 1° error in direction. A computer could
easily keep track of its motions. There is no uncertainty in the ball's
future! No chaos!

The situation is dramatically different when we have several balls
available for collisions on the table. The curvature of two colliding balls
produces an amplification of a factor of roughly ten in the directional
error. And the next time that the cue ball strikes another ball, its current
directional error is further multiplied by ten. After 100 collisions then,
an initial, ludicrously small error will have been multiplied by 10, one
hundred times. We now recall our elementary arithmetic. Every time
we multiply our error by 10 we move the decimal point one position
to the right. So after 100 collisions the decimal point has leapfrogged
to the right 100 jumps, which takes it over the backs of all the 99
zeros until the hundredth collision takes the decimal point over the 1
so that now, after about 2 min, the error in direction of the cue ball is
1°. And now we see how chaos is emerging from the shadows given
enough time. For after the next collision the error will be 10°, after the
next, 100°, and after the 103rd collision the directional error is 1000°.
But 360° take us once around a complete circle; so if we are 1000°
uncertain in direction we literally haven't a clue which way the cue ball
is moving. We don't even know whether it should be moving to the
left or the right, or up or down the table!

This is chaos of the most flagrant sort! An unbelievably small
uncertainty in initial conditions is amplified exponentially in a few
minutes to total inevitable uncertainty. And this amplification of the
error by 10 times 10 times 10 ... is what we mean by exponential
increase. Quite beautifully, our billiard ball scenario shows very
clearly how chaos lies hidden below the surface of a superficially linear
predictable world for a while until its time comes (after 100 collisions
in the billiard game) to show its unpredictable colours.

CHAPTER 12

TIME IN THE QUANTUM WORLD

Newton, Laplace and the vast majority of the great scientists of the classical era regarded time as an absolute quantity flowing in some determined way like a great river guided by the solid banks of reality. Einstein was to show in his special theory of relativity that in fact time is not absolute; there are many rivers of time, each flowing at its own speed. What Heisenberg and Schrödinger later discovered was that in addition the river banks were not as solid as Newton and Einstein thought. In their voyages through time, electrons and nuclei could shift from river to river, i.e. from future to future, in strange and ghostly quantum jumps.

Man who made sense of the mad microworld

For 4 days beginning March 1987, the quantum ghost of Erwin Schrödinger walked the passages and haunted the halls of Imperial College, London, as some of the great representatives of modern science gathered 100 years after his birth to pay homage to the memory of the man and his famous equation. The scientific stars were out to laud Schrödinger because, early in 1926, he published an extraordinary equation that codified much of the ghostly behaviour of the microworld. Schrödinger's equation became the password for entry into the asylum of electrons and nuclei—a world where common sense is nonsense, and where the unpredictable is the norm.

Schrödinger's work, together with the ideas of contemporaries such as Werner Heisenberg, Max Born, Pascual Jordan and Paul Dirac, utterly revolutionized our view of the world. Under their burning gaze, the too, too solid flesh of the world melted, not into a Shakespearean dew, but into an infinite sea of probabilistic waves. It is no wonder that Niels Bohr, the great Danish scientist, himself one of the pioneer explorers of the quantum world, would say that 'anyone who is not shocked by quantum mechanics has not understood it'.

The first 30 years of this century saw an intellectual upheaval without parallel in our search for understanding of the universe. First came relativity, which interwove space and time into a gravitationally warped tapestry and banished the concept of absolute simultaneity. Far more bafflingly came quantum mechanics, which shrouded the detailed happenings of the microworld in a fog of probability, and yet predicted the outcome of experiments with great accuracy.

The world view that Schrödinger helped overthrow was passed down to us from Galileo and, principally, Newton. This philosophy was perhaps best expressed by Pierre Laplace ('the Newton of France'), who (as we saw in chapter 11) was moved to imagine a super-intelligence, to which nothing would be uncertain, and both future and past would be present before its eyes.

The deterministic view of the world of particles dominated physics for 200 years as more and more mysteries fell before it. What led to its downfall was light. In 1900, the German physicist Max Planck discovered that, to explain the way in which a hot body absorbs and radiates light, he had to suppose that the process was jerky. It had none of the smoothness that one normally associated with the gentle rocking of waves that light was supposed to be made of. The energy transfers involved seemed to come in packages or quanta.

Planck, a brilliant but conservative man, had caught the first glimpse of the puzzling discontinuities of the quantum world. But he was no revolutionary at heart and failed to conclude that rays of light are composed of little packages of energy. It was left to Albert Einstein, a revolutionary's revolutionary, to do this 5 years later, in 1905. A key fact of quantum life was thus known, but far from understood: in some circumstances, light is best regarded as a wave, whereas in other situations (as Einstein had shown) light is far more usefully treated as a shower of minute particles.

Such a wave–particle duality is truly amazing, because the essence of a wave is that its energy spreads far and wide, like ripples on a lake. In contrast, particles are 'chunky things' that try to hang on to their energy. Even Einstein could not fully understand this schizoid behaviour, and he concentrated on developing his general theory of relativity. The next step to quantum understanding was taken in 1913 by Bohr, who took the ideas of Planck and Einstein to heart and applied them dramatically to the questions of the structure of atoms and how they emit light. Bohr's theory enjoyed some brilliant successes, but it juggled old ideas with new and didn't take the essential plunge into the phantom world of the quantum.

This move was made by the French physicist Louis de Broglie, who had the brilliantly simple idea that, if wave-like light often behaved as though it were composed of particles, then maybe objects that we usually think of as particles, such as electrons, could sometimes behave as though they were waves. de Broglie ingeniously showed how his electron waves could fit into Bohr's electron orbits in atoms. However, while de Broglie's flash of insight was brilliant, it lacked an adequate underlying theory for the behaviour of these mysterious waves that were to be associated with particles.

This was what was provided by Schrödinger, then a professor at the University of Zurich. What he came up with is known as Schrödinger's equation, one of the most celebrated equations in physics. It consists of a set of symbols resembling those used to describe waves on violin strings or those of oceans. But it is subtly different. Schrödinger showed that well known regularities in the light emitted from atoms were a natural outcome from solving his equation. It rapidly became obvious that his wave equation was the fundamental equation describing the microworld.

Just a few months earlier, first the 23-year-old German physicist Heisenberg and then the English theorist Dirac (nine months younger than Heisenberg), had developed apparently utterly different-looking theories of quantum behaviour that also seemed to work. Schrödinger quickly showed that all three theories were really the same. It was like having Darwin's *The Origin of Species* in three different translations— all told the same story.

The special beauty of Schrödinger's theory was that it was in a language well known to all practising physicists and was immediately applied successfully to many problems. But there remained a supreme irony with Schrödinger's wave equation. It could be said that the creator did not understand his own equation. There's no problem with waves on a violin string. It is the string that is waving up and down. It is the water that is bobbing up and down in lakes. So what is waving in the wave associated with electrons? Schrödinger thought it was the electron itself that was somehow smeared out into a wave. But this was soon seen to be impossible.

The correct explanation was proposed by the German physicist Born, with whom Heisenberg had been working. He said that the waves are waves of probability. When a moving electron is described by a wave drawn from Schrödinger's equation, the 'strength' of the wave at any point is simply the probability of finding the electron at that point! With this interpretation of Schrödinger's equation, the deterministic universe of Newton and Laplace collapsed in the microscopic world. According to Born, it is impossible, even in principle, to have a Laplacian super-intellect that could perfectly predict the future in the world of electrons and nuclei.

The world of the ultrasmall evidently reveals itself to us via probabilities. The evolution of these probabilities is, however, precisely described by Schrödinger's equation, which is like a mathematical crystal ball. The fortunes that it tells are those of electrons and protons and other inmates of the quantum asylum. A physicist wishing to forecast their future, in essence, sets the crystal ball into vibration by mathematically 'tapping' on it in a sort of Morse code, which tells the ball all that he or she knows about the present state of a system. In response, a series of possible futures appears in the ball, like a variety of possible endings of a play. Unable to predict the definite future, the crystal indicates the probable outcomes by listing the precise odds of each scenario.

Remarkable though it seems, while the vast majority of physicists have accepted the fundamentally uncertain nature of our knowledge of the microworld, three of the great quantum pioneers, Einstein, de Broglie and, yes, Schrödinger himself, never wholeheartedly went along with it. To express his unhappiness with the situation, Schrödinger invented the following thought experiment to illustrate just how 'unreal' the world of quantum mechanics is.

Schrödinger's cat

Imagine a live cat inside a closed metal box. Also in the box are a radioactive source and a Geiger counter, which can detect any particle emitted by the source. Thus, the microworld and the 'real' world are joined. The Geiger counter is adjusted so that, when such a particle is detected, it will trigger a hammer to smash a flask of poison gas, killing the cat instantly (figure 12.1). According to quantum theory, radioactivity is a random process described by rules of probability. Schrödinger, therefore, imagined that there is just the right amount of radioactivity in the box so that the probability that just one particle is detected by the Geiger counter in an hour is 50%.

Figure 12.1: *Schrödinger's cat.*

After the cat has been in the box for 1 h, the box is opened. Obviously, there is a 50% chance that the cat will be found alive. Anyone with common sense will surely say that, if the cat is found dead, it is because the Geiger counter detected a particle and triggered the release of the poisonous gas. Alternately, if the cat is found alive, the obvious explanation is that by chance no radioactive atom disintegrated; so the Geiger counter wasn't triggered to release the gas.

However, Schrödinger argued that the standard interpretation of quantum mechanics implied that, until the box was opened, and we actually looked inside, the cat was in a hybrid state of being alive and dead. Schrödinger thought this was absurd, as would the average person. While his equation was (and remains) brilliantly successful, he doubted that we understood its ultimate relationship to reality. Einstein described Schrödinger's cat paradox as the 'prettiest way' of illustrating his belief that quantum theory, although it is a fantastically successful tool, is not a complete description of reality.

Quantum mechanics explains radioactivity

In chapters (3) and (4) we saw how the discovery of the radioactivity of uranium revolutionized our understanding of the age of the earth and the time scale of our own emergence from 'Darwin's waiting room'. Now that we have met Schrödinger's equation, we are in a position to see how, in 1928, a whimsical young Russian physicist George Gamow used Schrödinger's equation to explain the random emission of alpha particles from radioactive nuclei such as uranium. Gamow illuminated in this way how time itself plays a fundamental role in alpha decay.

The phenomenon of radioactivity was well known but poorly understood when Gamow attacked the problem. In 1896, the French physicist Henri Becquerel had discovered that uranium compounds were spontaneously emitting a peculiar radiation which could fog photographic plates. Later work soon revealed that there were three components in the radiation, namely, gamma, beta and alpha rays, and all could be emitted from unstable nuclei. The gamma rays turned out to be a very high energy form of light. Beta rays were streams of electrons, while alpha rays were positively charged particles about 8000 times more massive than electrons.

Gamow set out to find a quantum explanation of how alpha particles could escape from the unstable uranium nucleus. He also wanted to discover why alpha decay seemed to be random. Given, say, 2 billion uranium-238 atoms, we know that after about 4.5 billion years (the age of the earth, coincidentally) one half of the uranium nuclei will have decayed, each by emitting an alpha particle. The randomness enters, however, because we don't know, and could not have predicted with certainty, which individual nuclei will have decayed. Because one half of any given number of uranium-238 nuclei always decay in 4.5 billion years, scientists call this period the half-life of uranium-238. (The '238' stands for the number of neutrons and protons in the nucleus in this particular form of uranium.)

In searching for an explanation of these features, Gamow needed a model of the nucleus and the potential escapee alpha particle within it. He found it in some work by Ernest Rutherford. This suggested to Gamow that the alpha particle could be regarded as though it were trapped within the deep crater of a volcano, its energy of motion being nowhere near enough to take it up to the top of the crater wall and to freedom down the outside flanks.

Before quantum explanations were accepted, physics theory said that such a situation would be completely stable. The alpha particle could never escape from the nucleus (the volcano) and uranium-238 would therefore not be radioactive. But it is radioactive. It can lose an alpha particle. Gamow decided to see whether the Jekyll-and-Hyde wave–particle duality of the quantum world would enable the alpha particle to escape through the volcano's walls. He set out to find from Schrödinger's equation the probability wave describing the alpha particle in the 'nuclear volcano'.

The remarkable result was that the wave was not quite totally confined to the crater. Schrödinger's equation said that an incredibly tiny but critical amount of the wave would leak through the volcano walls to the exterior! That is, on Born's interpretation of the waves, because the 'strength' of the probability is not zero outside the volcano (i.e. outside the nucleus) then there is a very small but finite chance that the alpha particle will at some stage be found outside as well.

Gamow's calculations indicated that something like a mere one hundred-trillion-trillion-trillionth of the wave's 'strength' existed outside the nucleus. The rest was inside. The Russian theorist interpreted this to mean that only once in one hundred trillion trillion trillion attempts, on average, would an alpha particle escape from a uranium-238 nucleus. At first sight, this number of required attempts seems so ridiculously large that one might well conclude that the nucleus would be effectively stable. However, when Gamow calculated how often the alpha particle tries to escape, he found another huge number, although smaller: one billion trillion times per second. At that rate, on average, half of the uranium nuclei in a deposit would have each lost an alpha particle in about 4 billion years, which accounts for the enormous half-life of uranium-238.

The number of attempts needed for an alpha particle to escape obviously depends on the height and thickness of the walls of the 'nuclear volcano'. These vary from element to element, and in the case of polonium-214 (see chapter 5) the alpha particle escapes so easily that the half-life for this polonium decay is less than the blink of an eye. This escape of an alpha particle from a force field prison because of the 'leakage' of the quantum wave through a barrier is called 'quantum tunnelling', and was also proposed simultaneously by Ronald Gurney and Edward Condon.

You might well wonder whether in the world visible to the naked eye, this ghostly passing through walls can occur? Gamow once calculated that the chance of a ball passing through a wooden block is so small that, if you began to write down the number of attempts needed for such a ghostly passage to occur, on average, the paper used would reach the region of the universe that is just visible through the approximately 508 cm Mt Palomar telescope by the time that you (or your descendants) arrived at the last digit!

Alice in quantum land

In addition to making several such brilliant contributions to theoretical physics, Gamow became a hugely successful writer of popular physics books. In the Gamovian spirit, therefore, we add a short fable about Alice in quantum land. For teenagers of all ages we examine the quaint behaviour of a quantum particle called alpha, which takes the form of a bee trapped in a prison called uranium. The story begins after Alice has entered a room only to have the door disappear after she shuts it. The only visible means of escape is a tiny mousehole.

Tears begin to appear in her eyes but, before a drop can roll down her cheek, Alice suddenly hears a curious buzzing sound that grows louder and louder. Wiping her eyes, she sees that the room is filled with bees,

Figure 12.2: *Alice might have escaped from quantum land by drinking and shrinking (top).   Instead, she used quantum tunnelling (bottom) (courtesy Richard Whyte).*

and, as they buzz around her head, that they are all dressed in beautiful brown-and-yellow pyjamas (figure 12.2). Stencilled on the back of each is the name ALPHA.

'Oh, thank heavens you're all here!' Alice cries, overjoyed to have company. 'Maybe one of you can help me to get out of here.' And then the most extraordinary thing happens; all the bees reply at the same time: 'Yes, buzz,' they chorus, 'I can, buzz, show you how to buzz out.' 'Oh, Dinah (her cat) will be pleased,' Alice says, 'but why are you all called ALPHA and why do you all speak at once? In fact, why do you all look identical?' The bees all buzz with laughter and explain, 'But there is only one ALPHA here. I'm a quantum bee and I 'bee-have' differently from ordinary bees.'

'Quantum bees,' Alice thinks aloud, 'quantum bees? I've heard of queen bees. Why, these are more puzzling than Cheshire cats. At least there was only one of them, and all it did was grin and disappear.' 'Oh, that's easy,' say the quantum bees and after zooming backwards and forwards across the room and bumping into the walls and ceiling many times, they suddenly disappear, all at the same time. 'Oh, dear,' says Alice. 'I didn't mean it badly. I think you're a wonderful bee. Won't you come back?' Then, suddenly, and to her joy, they are back, pyjamas flapping as they zoom around.

'If you listen to me,' say the bees, 'I'll tell you how to escape. It's an old quantum trick that was discovered by the first bee. She was called ALPHA PARTICLE and she lived about 15 billion years ago. She was trapped inside what the outside world calls a "nucleus" and she thought she was there for ever. But she flew backwards and forwards millions of times against the walls of the nucleus and, suddenly, although she didn't feel anything odd, she was outside! And every generation of ALPHA bees has carried on the secret.'

Alice can hardly believe it, 'And you mean that if I run backwards and forwards in the room, eventually I'll escape? Won't it hurt going through the wall?' 'No,' laugh the ALPHA bees. Alice closes her eyes and does as the bees described. She even makes a few buzzing sounds, 'just in case it helps'. Suddenly, the buzzing dies out and she opens her eyes. There is Dinah nuzzling against her leg—she is home in the 'real' world.

CHAPTER 13

IMPOSSIBLE THINGS BEFORE BREAKFAST

Said the White Queen to Alice in *Through the Looking Glass*, 'Why, sometimes I've believed as many as six impossible things before breakfast.' What you are about to read, before, after or during breakfast, however, is not impossible.

No material object can travel as fast as light. Light would take about 100 000 years to cross our galaxy. But, as mathematician Rudi Rucker has emphasized, you could cross it in a day, in principle. In fact, it is merely a problem in engineering and logistics (however difficult) to travel to any part of the universe in a day. Come to think of it, if light only travelled at 1600 km h^{-1} (instead of its actual 300 000 km s^{-1}), and you still could not equal its speed, you could still, in principle, visit any place in the universe in a day.

At this stage, you might well say, 'Wait a minute. I thought there were stars a billion light years from earth. That means light, the fastest-moving thing in the world, would take a billion years to reach them from earth. How then could something (such as someone in a rocketship) moving more slowly than light reach such stars in one day? Pass the cornflakes.'

The answer, in a word, is 'relativity'. For, in many ways, the special theory of relativity, pioneered by Hendrik Lorentz and Henri Poincaré and brought to full bloom by Albert Einstein and Hermann Minkowski, threw common sense out of the window. Said Einstein, 'Common sense is that layer of prejudice laid down in the mind before the age of 18.'

Einstein based this 'outrageous' theory on two postulates: one seemingly innocuous and the other much more bizarre. The inoffensive postulate merely reflects the experience of anyone who has flown in a large modern jetliner. If the plane is flying at a steady speed, not being accelerated by the pilot or thrown around by turbulence, and if all the window blinds are drawn, you can't tell that you're moving. The stewardess, for instance, pours your coffee exactly as she would on the ground. She doesn't have to stand at the front of the plane pouring while the plane whips your cup under the fluid at 1000 km h^{-1}. If your plane were flying at 96 000 km h^{-1} she'd still pour for you in the old-fashioned way. In the theory of relativity, this experience is generalized into Einstein's first postulate to say that no experiment of any kind, whatever its degree of sophistication, carried out inside a non-accelerating compartment can tell you whether that compartment is moving. As far as physics is concerned, the people in the smoothly flying jet have just as much right to consider themselves stationary as have their families left behind standing on the ground. To the jet passengers, it is the people on the ground who are moving at 1000 km h^{-1} in the opposite direction.

The second postulate is much more odd and was bound to cause something in our common sense to recoil. It says that, regardless of whether you rush headlong at a light beam, or try to outpace it, you will always find light passes you at the same speed: 300 000 km s^{-1}.

This is obviously absurd. It doesn't make sense, at first sight. We all know that, if we are driving along the road at 60 km h^{-1}, and pass a car going in the opposite direction at the same speed, then the other car passes us at 120 km h^{-1}; we overtake a car being driven at 40 km h^{-1} and we shall pass it at a mere 20 km h^{-1}.

Surely then, if we rocket towards a star at 99% of the speed of light, won't the star's light pass us at almost twice the usual speed? Or, if we shot directly away from the star at 99% of the speed of light wouldn't we almost leave the light behind? Wouldn't the starlight only overtake us at 1% of the usual speed? The answer in each case is no. The light will always pass at 300 000 km s^{-1}. The answer is truly in defiance of one's ordinary common sense and inevitably means we have to change some of our prejudices.

Since our paradoxes all seem to have revolved around the word 'velocity', it is obvious (if you're Einstein) that you have to say, 'There must be something wrong about our concepts of velocity.' But what is velocity? It seems so simple, so foolproof. It's merely the distance that you travel in a certain time, such as 30 km h^{-1}. What could be misunderstood about that?

Since velocity involves only the two concepts, distance and time, our notion of distance and/or time must be wrong, Einstein concluded after puzzling over these seemingly contradictory phenomena in nature. He deduced, from his two postulates, that what we consider to be the length of anything depends on its velocity relative to us. Thus a rocket flashing past us at about 87% of the speed of light would seem only half as long to us as when it was stationary!

Having thus demolished the idea of an absolute length, Einstein then laid to rest absolute time. He pointed out that, not only would we find the moving rocket's length to be less than when it was stationary with respect to us, but also we would consider that any clocks on board the moving rocket were ticking much more slowly than ours even though, before the rocket took off, its clocks kept perfect time with ours.

But that's only half the story. Because, by the first postulate of relativity, the astronauts in the rocket, after all the acceleration had ceased, can with full justification consider themselves to be at rest, if they have a mind to. They, looking at the earth, will see it hurtling past them at 87% of the speed of light and will conclude that the earth has shrunk by a factor of 2 in the direction of its motion. The astronauts will also conclude that all the earthly clocks are ticking too slowly. In contrast, they will say their rocket has not shrunk in any way. It's just as roomy as it was on earth. And they don't consider that their clocks run differently than they did before blast-off.

Common sense should now tell you to say, 'But which of these viewpoints is right? Are the astronauts right, or are we right?' And relativity theory will answer, 'You're asking the wrong questions. Because both are right!' According to relativity theory, 'All non-accelerating observers are born equal'; in a sense, we all carry our own time and space with us.

If we come back now to the possibility of a person travelling the width of the Milky Way in a day, an idea mentioned at the beginning of the chapter, we can see now how he or she can cross the galaxy in a day, while light 'would take 100 000 years'. It's all a matter of 'whose day?' and whose '100 000 years'. For the humans who opted to stay at home and to watch the astronauts travel to the other side of the galaxy, the trip would, in fact, take longer than 100 000 years, because humans cannot travel as quickly as light. But these 'sticks-in-the-mud' would agree that 'for the astronauts' only one day will pass on the voyage. Because, say the homebodies, the astronauts' clocks (and their hearts) are all slowed down while they are travelling at almost the speed of light.

'But, but...' you say, 'I thought that by the first postulate of relativity, the astronauts don't consider themselves to be aging any more slowly than when they were in training school on earth. So, from their viewpoint, their aging process should be unchanged and they should be dead long before they reach their goal.' To this, the relativist replies, 'Yes, the astronaut does age normally. But, in considering himself at rest, he (the astronaut) sees the galaxy hurtling past at almost the speed of light and therefore he sees the width of the galaxy shrunk to almost trivial size; so naturally it flashes past him in just one day!'

An illustration of this effect is happening all the time in nature. A tiny particle called a meson is produced in the earth's atmosphere at high altitudes; yet, despite existing for only the merest fraction of a second, it can reach the ground. Instead of only travelling a few hundred metres through the atmosphere before disintegrating, as one would have calculated through common sense, the particle 'sees' the atmosphere rushing past itself and therefore severely contracted, and it is easily struck by the ground before it 'dies'.

If the relativistic challenge to your common sense leaves you dismayed (instead of, as I hope, exhilarated) you won't be alone. Sir John Squire reacted this way in reply to Alexander Pope's famous epitaph for Isaac Newton ('Nature, and Nature's laws, lay hid in night: God said, "Let Newton be!" and all was light'). Sir John wrote,

> It did not last: the Devil howling 'ho!
> Let Einstein be!' restored the status quo.

Perhaps Einstein did sometimes believe six impossible things before breakfast!

Back to the Future

We're now ready to meet the mind-bending twin paradox which has led to an enormous amount of debate and yet which was not considered at all paradoxical by Einstein or the vast majority of physicists.

Suppose that our intrepid astronaut, John, had an identical twin, Charles, who stayed behind on earth. This time, imagine that John was travelling at almost the speed of light, so that he reached in 1 day (on his own clocks) the vicinity of a mysterious star which earthlings have long considered to be 0.5 million light years away. Then suppose that John sees what he wanted to see and returns to the earth in one more day of his own time.

Obviously (?), as far as John is concerned, he returns home 2 days older. As he knew would be the case, he finds that over 1 million years have elapsed on earth since he left. His twin brother Charles will have been dead for over 1 million years, while he John is only 2 days older than when he left!

How can we possibly explain this? Before Einstein it would, of course, have seemed impossible. Einstein would say, 'It's obvious.' The star is 0.5 million light years away. So light would take 0.5 million years to reach it from earth and 0.5 million years to travel from the star back to earth. That is, light would take 1 million of Charles' years to make the stellar round-trip. Since John's rocket is not quite travelling at the speed of light, Charles says it must take slightly longer to make such a trip than does light and so John will return to earth just over 1 million of Charles' years after he left.

So there you have it. John has somehow in two of his days travelled 1 million years into the earth's future. Evidently by travelling for 2 days at a speed even closer to that of light, to and from a much more distant region of the universe, John could return over 100 million years from now and see that new oceans have opened wide and old oceans have closed, and strange new creatures are swarming on its surface!

But now the Devil howls 'ho!' and plays his own advocate. He says, 'Wait a minute (of my Devilish time). There's a paradox here! I thought you said earlier, Einstein, that all observers are equivalent. I say then, suppose that we look at the scenario as though the rocket and John didn't go anywhere, but the earth and Charles hurtled off for 2 days of their time and met John on their return. I thought you could look at things just as correctly this way. And, if so, when the twins re-united, this time Charles would be only 2 days older whereas John now would be 1 million years older. Surely we have a paradox here. They can't both be older than their twin!'

To which Einstein would reply that the Devil has made a mistake. The Devil has assumed a perfect symmetry between the two scenarios, which is wrong. John and Charles have undergone quite different physical experiences, for John undoubtedly experienced considerable accelerations in leaving the earth and reaching almost the speed of light, and in turning around at the midpoint of his journey. In contrast, Charles was blissfully unaware of any such effects on himself. So the Devilish argument, which was based entirely on symmetry, is invalid. John will return to the future earth!

CHAPTER 14

MUCH ADO ABOUT CANNONBALLS (AND DEMOCRACIES); LAST EXIT TO PISA, NEXT EXIT BLACK HOLES

Galileo's experiment, in which he purportedly dropped cannonballs and lighter objects from the leaning tower of Pisa, will always occupy a special place in the hearts of scientists. The results deduced from it were key components of Isaac Newton's law of gravity in the seventeenth century and of Albert Einstein's general theory of relativity. Each of these towering geniuses realized the critical nature of Galileo's conclusions (bodies of different masses fall from rest at the same rate, in a vacuum, towards the earth). However, they built them into conceptual frameworks of staggeringly different types, even though they were both trying to explain the same universe. Einstein was even ultimately able to link time and gravity from these considerations.

Newton's thinking was 'force oriented.' He realized that, if you give a skater a push, you give him an acceleration. And, if you use the same amount of push on a skater twice as large, he will suffer only half the acceleration. Applying this idea to Galileo's result, Newton realized that, if large objects fall towards the earth at the same rate as small objects, it must be because the earth pulled harder on the more reluctant large objects. So, on formulating his celebrated gravitational theory, Newton included precisely this factor of mass dependence of the gravitational force. 'Gravitational democracy' (all bodies fall from rest at the same rate in a vacuum) was thus built into

Newton's equations and for more than 200 years everybody accepted this. Brilliant mathematicians and astronomers were able to describe the motions of virtually all the known members of the solar system in its terms.

In 1907, 2 years after publishing his extraordinary paper on the special theory of relativity, we can imagine that the 27-year-old Albert Einstein was putting aside consideration of the successes of Newton's gravitational theory, and thinking again about Galileo's experiment! And he was, just like Newton, thinking about forces, masses and accelerations. However, as these concepts moved around in his brain, patterns of a very different nature emerged than had appeared in Newton's mind. The profound importance of gravitational democracy of course still stood out like a rock in the sea of ideas. But then a different principle of democracy started to emerge above the waves of Einstein's subconsciousness.

In keeping with Einstein's mode of thought, we shall call this 'passenger democracy'. Imagine, together with Einstein, a group of passengers on a train. When the train is moving at constant speed, everyone is sitting relaxed in his or her seat. In fact, if the train's movement were perfectly smooth, and a slim girl in the corner didn't look out of the window, she wouldn't know she was moving. Now suppose that the train driver spots an emergency and slams on the brakes. As we all know, all the passengers will be hurled forwards immediately. But what Einstein seized on was the completely democratic nature of the whole experience. That is, he focused on the well known fact that every passenger who was free to move would have been thrown forwards through the air at exactly the same rate. The fat man and the slim girl sitting next to him would hurtle forward in parallel! That's democracy.

Einstein now put these two democratic principles together in his head. Cannonballs and feathers fall down together in a gravitational field (in a vacuum). And fat and thin passengers (and anything else freely movable) 'fall' forwards together in a steadily braking train. This unique thinker then questioned whether these two democratic principles are really different or intimately related i.e. whether they are two sides of one coin.

Einstein concluded that they are related and he codified this in his celebrated principle of equivalence. He reasoned that there are two equally good ways of interpreting the events that transpired on the train that day. First, there is the obvious way, namely that the train suddenly began to decelerate and all the passengers were flung forwards. Alternatively, however, we could look at things from the point of view of the young woman passenger. She had been sleeping soundly for an hour and had wakened only a few minutes before the emergency. All the blinds were drawn, and the young woman naturally presumed the train had stopped while she was asleep. After all, in an unaccelerated compartment, as we all know from experience, there is no sign of any motion.

In her description of the events afterwards, she would say, 'The train was standing there motionless, when suddenly we all fell forwards together! It was just as though the engineer had taken on a load of coal at the front of the train (relativity grew up in the era of coal-fired locomotives) the size of the earth and we all fell down horizontally towards it together because of its gravitational field.'

To Einstein, the woman's viewpoint was perfectly valid. For him, it was equally acceptable to describe the events as occurring with respect to a decelerating train, or with respect to a stationary train in a horizontal gravitational field. Einstein then realized that, if he calculated some of the things that go on in an accelerating (or decelerating) train, he could then turn around and say that these are exactly the things that would happen in a stationary train in a horizontal gravitational field. That is, he gained totally new insights into the nature of the gravitational field! This led him to realize that paths of light would be 'bent' by gravity, and that clocks would run at different rates in gravitational fields (see p 118). He realized that it implied that the space–time structure of the universe itself would be warped by matter.

Finally, it led to the removal of the Newtonian idea of force from gravity. The earth moves around the sun not because of the sun's gravitational pull but because it's natural to move like that in the 'warped world' near the sun. So that's why Galileo's experiment has long been a big deal, even though some historians contend that he probably didn't really carry it out.

Bleak holes–time's grave

In our discussion of gravitational democracy we saw a beautiful example of Einstein's unmatched ability to look into the heart of extreme complexity and see simplicity itself. It was as though he were gifted with a special child-like naivety which prompted him to ask the simplest of questions—questions overlooked by other gifted, but more mathematically oriented, physicists. Einstein's first step in solving a problem always seems to have been to find the most easy analogy or model and, if possible, to deduce solutions from this simple model with the simplest mathematics appropriate to the situation. The best illustration of this was his unveiling of the principles of the special theory of relativity. The physics genius Hendrik Lorentz and the great mathematician Henri Poincaré had found most of the elements of this theory, and yet failed to 'see' it and to present it in the exquisitely simple and complete way of Einstein.

Another example of the power of Einstein's 'naive' approach is the way that he interwove time and gravity, concepts which had previously been regarded as totally independent, as he struggled to develop the insights he would need to formulate the general theory of relativity and gravity. When we talked about how one could (at least in principle) cross the galaxy (or the universe) in a day, or when we examined the twin paradox, we saw how Einstein had overturned our views of time. In particular, he pointed out that a clock moving with respect to us ticks slowly from our point of view. Now, as he wrestled with the thoughts of gravity and time, he was struck by a resemblance between gravity and that other force we're all familiar with—centrifugal force.

We remember from our discussion of 'passenger' democracy (p 114) that if we are in a train that abruptly slows down, all of us experience a force which flings us forwards democratically. That is, fat men and slim women 'fall' forwards together, just as cannonballs and feathers fall downwards together (in a vacuum) in a gravitational field. Therefore Einstein said that this force of democracy that flings us forwards because of a change in motion of the train is indistinguishable from gravity in its effect. Now, let's look at centrifugal force.

Suppose that our train is moving at a constant speed when it comes to a bend in the track. As the train takes the turn, we are all (fat and thin) flung together against the compartment wall towards the outside of the curve. Everyone knows that the force that flung us sideways is called centrifugal force. It's just the force that you always experience (continuously) as you sit on an animal on a child's roundabout (merry-go-round). But now in came Einstein's child-like genius. In essence, he reasoned that since the original democratic force came about because of a change in the train's motion (its braking, in chapter 13) and since the centrifugal force (which seemed to be pretty democratic in the way it flung us all outwards) emerged because of another change in the train's motion (its veering round the bend), then *maybe the centrifugal force can also be looked on in some way as gravity!*

And this beautifully simple thought (once you've thought of it) immediately suggested another ludicrously simple thought experiment that would immediately reveal to Einstein that gravitational fields affect the ticking rates of clocks! That such a few breathtakingly simple thoughts could achieve this insight is one of the intellectual wonders of the world; Einstein had seen how clocks and, therefore, time itself are intimately involved with gravity! And, thanks to Einstein, it's as easy to understand as riding on a 'roundabout'.

Imagine, with Einstein, a simple roundabout. It's rotating smoothly, with lots of children laughing and squealing as the inevitable music plays. One of the children, a boy, sitting still on a horse fixed near the edge of the roundabout, is wearing a watch. A young girl sitting on a giraffe near the centre is wearing an identical watch. By chance, the watches were bought on the same day, in the same store and were keeping identical time in the store. Einstein is standing on *terra firma* watching all this and looking for clues about a theory that will describe the evolution of the universe! And then he spots it—an extraordinary effect that he should have thought of earlier (as we mortals say). Because of his rotation with the merry-go-round, the boy's watch is moving relative to Einstein and so, as described by the special theory of relativity, it is running slowly! The girl's clock is also moving, but much more slowly past Einstein because she's nearer the centre of rotation; so her watch is only slightly affected and is only running a little slow relative to Einstein's watch. Einstein realizes that someone on the roundabout, at its very centre with a watch, is not moving to or from him at all. His is not a moving clock, and therefore it reads the same as Einstein's watch. Therefore, Einstein says that it is obvious that the central observer will see exactly the same effect as he (Einstein); in other words, to observers on the roundabout, watches at different positions on the rotating roundabout will be running at different rates.

And now comes the pay-off punch. Einstein asked himself what conceivable phenomenon or force can be causing this variation between the watches seen by the people on the roundabout? Why, the *centrifugal force*, of course! What else is there? What's more, it varies in just the right way with position on the merry-go-round. It's greatest farthest from the centre of rotation and zero at the centre, just like the 'irregularities' on the watches. But he had suggested earlier that *centrifugal force can be looked on in some way as gravity*! Einstein therefore immediately leapt from the roundabout to the stars with the words, 'Thus on our circular disc (roundabout), or, to make the case more general, *in every gravitational field, a clock will go more or less quickly, according to the position in which the clock is situated (at rest).*'

But since time is what is measured by clocks, then the flow of time is affected by gravity! In a comparable tour de force of simple-minded thinking, Einstein then imagined what happened when a set of rulers were laid down around the circumference of the roundabout and realized that Euclid's geometry no longer held sway for the rotating children (for them there would not be 180° in a triangle) and therefore that gravity warped not only time, but also space. In 1916, Einstein revealed that it was precisely thoughts such as the above which led him to the enormously successful (and very mathematical) general theory of relativity. In this theory, space and time are coupled and it is usual to say that space–time is warped by gravity so that it is best described by non-Euclidean geometry.

Now that we are familiar with the slowing down of time by gravity, we are in a position to look at some strange time happenings near black holes. But first we must meet black holes.

Time to meet black holes

While it has been necessary to wait until the second part of the twentieth century for the apparent observation of black holes, these strange objects made their appearance in the minds of humans almost 200 years earlier.

In 1783, the Rector of Thornhill, Yorkshire, John Michell began to think about the old adage that what goes up must come down. We all know that, if we throw up a ball, it eventually stops and falls back down to us. The faster we throw the ball, the farther the ball climbs before it falls back.

It is natural to wonder whether the ball could be thrown so fast that it will never fall back. And the answer to this is 'yes'. If the ball is thrown upwards at about 11 km s^{-1} or faster, it will never fall back down (the slowing effect of the earth's atmosphere is ignored here); it will just go on for ever. This critical value of the ball's speed is known as the escape velocity of the earth.

What Michell essentially then realized was that, if the earth could be squeezed between the palms of some passing giant's hands, so that its radius was say, divided by 16, its escape velocity would be quadrupled to 44 km s^{-1}. But instead of stopping there, he took a remarkable leap and, in effect, concluded that, if the Earth were squeezed to about the size of a ping-pong ball, the escape velocity of this mini-planet would be greater than the speed of light.

If that happened, even light would not be fast enough to escape the earth. To the man in the moon, then, the earth would have become invisible. It would be what we now call a black hole. Not only did the remarkable rector think of black holes, he also, amazingly, worked out the correct formula for calculating a black hole's radius from its mass. Of course, he was lucky to get the right answer, a trait of genius, because there was no evidence whatsoever at the time that light was affected, like a ball, by gravity. Also, he was using Newton's theory of gravity instead of the approximation much nearer to the truth found by Einstein long after Michell's death.

Michell's formula for calculating a black hole's radius is extremely simple; the radius of a black hole measured in million-billion-billionths of a millimetre is approximately equal to the number of kilograms of mass of the object which has been black-holed.   Take Oxford mathematician Lewis Carroll's Alice. After Alice had fallen down the rabbit hole, she drank from a bottle labelled 'Drink Me'. The magical fluid made Alice fold up like a telescope until she was only about 25 cm tall. Then, in an extraordinary passage, Alice describes her fears of what sounds to modern ears like a collapse into a black hole. Says Alice to herself, 'For it [her shrinking] might end, you know, in my going out altogether, like a candle. I wonder what I should be like then?' The story goes on: 'And she tried to fancy what the flame of a candle looks like after the candle is blown out, for she could not remember ever having seen such a thing.' (figure 14.1).

Figure 14.1: *Alice waxes philosophical (courtesy Richard Whyte).*

If that's not the most beautiful description of the collapse of a dying star into a black hole, I don't know what is. Yet, so far as I know, Lewis Carroll had never heard of such conceivable astronomical oddities (nor had Alice). At any rate, it's very easy for us, using Michell's formula, to see at what point Alice's fears would have come only too true. If Alice's mass were, say, 30 kg, then by Michell's recipe, if she drank enough of the mysterious potion to shrink to about 30 million-billion-billionths of a millimetre (while preserving her mass), she would indeed have gone out altogether, 'like a candle'. It is apparently a fluke that Alice stopped short at 25 cm and thus preserved a literary masterpiece for posterity.

But if an Oxford mathematician of literary genius gave us the wittiest unwitting description of black hole formation, a Cambridge astronomer of world renown, Sir Arthur Eddington, gave the most ironically misplaced dismissal of these celestial phantoms.   Sir Arthur, an unsurpassed popular science writer as well as a great theorist, was commenting in 1935 on a theory developed by the outstanding Indian mathematician Subrahmanyan Chandrasekhar. The latter had been studying the gravitational collapse of stars after their nuclear fuels had run out. He had concluded using Einstein's gravitational theory that, while stars of small mass would collapse into so-called white dwarfs, larger stars would not reach some final state. As far as he could tell, they would go on collapsing. Chandrasekhar did not, however, conclude that in fact these stars would end up as black holes. He merely said, 'A star of large mass cannot pass into a white dwarf stage and one is left speculating (like Alice) on other possibilities.'

Sir Arthur said, when explaining the meaning of Chandrasekhar's theory, 'The star has to go on radiating and radiating and contracting and contracting until, I suppose, it gets down to a few kilometres radius when gravity becomes strong enough to hold the radiation and the star can at last find peace.' Chandrasekhar has since pointed out that had Sir Arthur stopped there he would now be getting credit for being 'the first (modern scientist) to predict the occurrence of black holes.' But the formidable Sir Arthur could not resist adding, 'I felt driven to the conclusion that this was almost a *reductio ad absurdum* of the relativistic degeneracy formula. Various accidents may intervene to save the star, but I want more protection than that. I think that there should be a law of nature to prevent the star from behaving in this absurd way.'

Unfortunately for him, Sir Arthur's combativeness and penchant for the witty remark led him astray on that occasion. Also, according to his biographer K C Wali, it probably discouraged Chandrasekhar (who was then a rising young man in contrast with the English superstar) from pursuing his theories and discovering black holes. It was, in fact, left to J Robert Oppenheimer and his co-workers G M Volkoff and H Snyder to track the collapsing stars to their resting places and, as a result, uncover black holes and neutron stars, another form of collapsed star.

Black holes and twins

Earlier we saw how after a space traveller returned to earth following a high-speed trip among the stars, he found that 1 million years had elapsed on earth while he was away. His twin who stayed behind on earth, had of course been dead for 1 million years! Now we imagine a scenario in which twins set off together in a rocket, seeking close-up information from a recently discovered black hole.

After a trip lasting, for them, a couple of years the space-ship nears the black hole and uses its engines to inject itself into an orbit around the hole. One morning twin Mary wakes up to find a note from twin John saying, 'Mary, I'm sorry, but I had to do it. I got this overwhelming urge to plunge into the black hole in my mini-craft and actually experience the ultimate. I'll keep sending light flashes back to you to keep in touch, as long as I can. Hoping we meet again some where in space–time; I love you, John.'

In horror, Mary, who had always understood general relativity better than John, stared out of the porthole toward the black hole. At first she could see nothing except background stars, but then she picked up John's regular light flashes merrily twinkling. And then the distorting effect of gravity on time began to manifest itself. For Mary didn't see John's signals for only a while before he vanished into the black hole and none of his signals could escape any longer. She found instead that light flashes from John kept on and on arriving, although they got very slowly weaker and the interval between flashes got gradually longer. And so it went on until Mary died of old age and a broken heart in the palatial spacecraft. Her children, who were born and grew up on board and their descendants, also kept watch for Uncle John but had no better luck than Mary. Ever more weakly and ever more delayed, but never stopping entirely, John's signals kept coming in from just above the black hole's surface. In fact, from the viewpoint of the orbiting spacecraft, John takes an infinite time to fall into the black hole!

The explanation for the mother craft observers is simple, if we think back to Einstein's roundabout. As John falls nearer and nearer to the black hole, its gravitational field gets stronger and stronger and therefore, in Mary's opinion, the gravity slows down the flow of his time. So the intervals between his light flashes increase steadily, and the oscillations in his light flashes slow down from flash to flash and transport less and less energy, until the time interval between flashes becomes infinite! From the orbiting craft's viewpoint, John never falls into the black hole.

John sees things totally differently! Instead of the life imprisonment in an ever-falling coffin that Mary perceives him suffering, John unfortunately rapidly meets with capital punishment. Because he is falling freely, he feels that he is weightless and therefore not even in a gravity field. Think of the astronauts that we've all seen on television floating around weightlessly in their earth-orbiting capsules. Despite being near the earth and its strong gravity field, because they are freely falling (around the earth, its true, but they are still falling always towards it; otherwise they'd shoot off into space) they 'feel' no gravity. So, since John feels no gravity, neither does his watch and he experiences no gravitational slowing of time. He therefore is aware of falling ever faster towards the black hole and very quickly crosses its border (its so-called event horizon)! The capital punishment is meted out even before he enters the black hole, because enormous tidal forces are exerted on him by the hole and he is quickly torn apart.

Just as before (pp 111 and 112), the twins experience totally different worlds, in which time flows completely and utterly differently for each! There is no such thing as a universal absolute time in the world that all observers can tune into 'from time to time' to synchronize their watches, Isaac Newton to the contrary notwithstanding.

CHAPTER 15

THE ARROW OF TIME

O! call back yesterday, bid time return
Shakespeare, *Richard II*, III, ii, 69.

Out, out, brief candle!
Life's but a walking shadow, a poor player,
That struts and frets his hour upon the stage,
And then is heard no more;...
Shakespeare, *Macbeth*, V, 16.

We can all call back yesterday, bid time return, but it won't make any difference! Our yesterdays are all gone, never to return. The poet and scientist agree; every one of us will strut and fret our hour upon the stage, and then we'll be heard no more. Time flows onwards, never stopping and never turning back. And in this very time, we shall all eventually grow old and lose our powers. We shall fall victim to the operations of the second law of thermodynamics which says that, as time rolls by, the disorder in the universe will steadily increase. To stay alive we have to maintain our extraordinary level of order (organization). Our failure to do this brings death.

That there is a quantity which steadily increases with time and which could therefore act as a sign post in the universe for the direction of the flow of its time was discovered in the nineteenth century by Rudolph Clausius, Lord Kelvin, Herman von Helmholtz and Ludwig Boltzmann. Clausius named it entropy, and Boltzmann showed that it was a measure of the degree of disorder of a system, whether one was discussing the expansion of steam in a steam-engine, the flow of heat out of the earth, or the behaviour of the universe.

124

If you were shown two perfect detailed pictures of the universe taken at two different moments in time you could always (in principle) decide which was taken first by calculating the entropy represented in each picture. The image showing the greater entropy represents the later universe. This capacity of entropy to point unerringly in the direction of time's flow, was called 'the arrow of time' by Sir Arthur Eddington. Ironically, while entropy is a subtle, even controversial concept as we shall see, we all have an innate ability to recognize its presence and its inexorable increase in nature. We all see it and recognize it every day in the tendency for things to wear out or for things to break far more easily than they can be repaired. Roofs leak and paint peels without you having to lift a finger to help the increasing dilapidation. In contrast, to repair your house requires intelligently directed (ordered) effort.

If we watch a movie, we can all tell within seconds whether or not it's being shown backwards. If we saw broken glass shards suddenly jump off the sidewalk and hurtle in beautifully choreographed orbits into a window frame and come to rest as a perfect pane of unbroken glass, and if the pane of glass threw a stone down into the outstretched hand of a grinning mischievous boy, we would all know that the film was being projected backwards!

If we saw the spray and foam at the foot of Niagara Falls re-united into more coherent droplets and these began to rise up the face of the falls, uniting into an upward flow of water back over the lip of the falls, we would realize that we were watching a movie running backwards. If we saw a huge dark mushroom cloud contract and shrink downwards into a collapsing fireball, and saw a devastated city spontaneously resurrect itself, with the help of a flaming inferno and gale force winds, we would know we were watching a film of the nuclear bombing of Hiroshima (or Nagasaki) in reversed time.

In all these cases we would spot things that were unnatural and unreal. Stones break windows. Water falls in waterfalls. Bombs destroy cities. And yet energetically there was nothing wrong or forbidden about any of the events in the movies. That most basic of physics laws, the conservation of energy, was never violated. The total energy at the beginning of the scenes shown exactly equalled the total energy at the end. The fundamental laws of motion in physics all work equally well with time running backwards or forwards; so none of the motions in the back-to-front movies was breaking Newton's or Schrödinger's laws. So why do backwards films look peculiar or, the same thing, why don't these funny things happen in real life if it wouldn't be violating any rules? Why doesn't a kettle freeze when you put it on the stove? Why does disorder increase in the direction in which time flows?

At this point, you might interject that, if disorder (entropy) is inevitably always increasing, how could we evolve and exist as highly ordered structures. Doesn't our very existence show that there are exceptions to the rule that the entropy of a system always increases? The answer is *No*! No living system is closed. Certainly the evolution and continuing existence of life on this planet represents an increase in order with time for those molecules involved in all our bodies and plants. But careful studies show that always associated with this sequestering and ordering of molecules is an accompanying production and loss of heat to the environment which always results in a disordering of our surroundings which is greater than our ordering. That is the total disorder, or entropy, in the whole system, life plus non-living environment, keeps increasing as time goes by.

So we are brought back again to the same question: why does disorder (entropy) steadily increase in the universe? We should admit at once that there is no universal agreement among physicists as to the answer to this question! A large group, epitomized by cosmologists such as Fred Hoyle, Roger Penrose and Stephen Hawking, say that it is all explained on cosmological grounds (perhaps not surprisingly, demonstrating some evidence of vested interests). In opposition to this, Peter Coveney and Roger Highfield in their book *The Arrow of Time* quote British theorist Peter Landsberg, approvingly, as saying, 'It seems an odd procedure to "explain" everyday occurrences, such as the diffusion of milk into coffee, by means of theories of the universe which are themselves less firmly established than the phenomena to be explained. Most people believe in explaining one set of things in terms of others about which they are more certain and the explanation of normal irreversible phenomena in terms of cosmological expansion is not in this category.' Coveney and Highfield then go on at this point in their book apparently to give up on finding an explanation of the 'arrow of time'. They write, 'Instead, we appeal to the second law (of thermodynamics) as a phenomenological fact of nature to select evolution to future equilibrium in accordance with what we observe. By this simple yet enormously significant step, the second law's 'arrow of time' is incorporated into the structure of dynamics, which thereby undergoes a quite radical alteration. Such a resolution of the irreversibility paradox, which puts the second law and mechanics on an equal footing, is quite different from the unsuccessful route taken by those who wish to see mechanics explain thermodynamics in the time-honoured reductionist spirit.' I take this to mean that entropy increases with time because entropy increases with time. It is hardly surprising that, when Claude Shannon was contemplating choosing the word 'entropy' as the measure of information in a message in his new 'information theory', the renowned mathematician John von Neumann said encouragingly to him, 'It will give you a great edge in debates because nobody really knows what entropy is anyway.'

In essence, the cosmological argument to explain the 'arrow of time', the steady increase of disorder in the universe, is simple, but it comes in two parts. We shall give the outline of the complete argument first, without justifying either of the steps so that the reader can appreciate the simplicity of the logic. Only then will we examine the justification for the two steps taken. *Step one* says that, if an isolated system is in a highly ordered (low-entropy) state it will inevitably evolve into a more disordered (higher-entropy) state. *Step two* follows with the assumption that the universe was formed at the Big Bang in a highly ordered state; so by step one it is naturally evolving into a more disordered state, i.e. the entropy of the universe is increasing. That's it.

Step one is easy to justify; as we said earlier in the chapter, the evidence is certainly all round us that it's true. All highly organized structures break down and decay. But how do we explain this in a more convincing abstract way? This was originally done by Boltzmann in his foundation of the subject of statistical mechanics. In our version we adopt the usual jigsaw puzzle analogy. When any such puzzle has been assembled, it's in a highly ordered state, and there's only one way of arranging the pieces to achieve this state. If we now shake up such an assembled puzzle in its box, the picture will gradually break up into more and more disconnected pieces. Eventually all the individual pieces will have been separated from each other and they will be randomly distributed in the box. Further shaking won't reassemble the puzzle! It merely changes one random heap of pieces into another random heap. The jigsaw puzzle has modelled the evolution of our universe. It began in a highly ordered state (with the completed picture) and gradually became more and more disordered. Eventually it (the puzzle in the box) reached a maximum degree of disorder, its entropy became a maximum and then we say the equivalent of thermal equilibrium was reached. If the universe keeps expanding long enough this could happen to it (see, for example, our discussion of Stephen Hawking's ideas a little later).

The jigsaw puzzle became more and more disordered because there are many, many more ways for a system to become disordered than there are for it to order itself. We can best see this by comparing the two end states: the initial highly ordered assembled puzzle versus the final totally randomized heap in the box. As we said above, there is only one arrangement of the pieces which completes the puzzle, whereas there is an enormous number of equally randomly distributed piles of disconnected loose pieces. Boltzmann pointed out that, if a system has a chance to explore all its various possible states, it's clearly going to end up in the highly disordered maximum-entropy state of thermal equilibrium simply because there are so many equivalent representatives of that state waiting to be found by the evolving system.

If, then, step one of our cosmological argument for increasing entropy can be reasonably justified by such probabilistic arguments, step two (the assumption that the universe began in a highly ordered state) is more difficult to explain in detail. While there are many indications that the universe began in a very smooth, highly organized (and therefore low-entropy) state, exactly what value (of order or entropy) that it had is unclear and depends on the model used for the Big Bang. Penrose, who elaborates on a cosmological–statistical argument such as the above in his book *The Emperor's New Mind*, suggests that something (which he does not yet understand) in the initial geometry of space–time at the Big Bang, causes the required high degree of order in the beginning.

If such speculations are correct, then, extraordinary as it may seem, water does boil on a hot stove, and milk does diffuse in coffee, because of the conditions in the Big Bang billions of years ago. Every time you sip a coffee you should meditate on that!

Entropy, the Big Bang and the Big Crunch

The grave difficulties in understanding entropy in its cosmological context can be further highlighted by the different behaviours of the 'arrow of time' painted by Penrose and Hawking in their recent books when the possible recollapse of the universe is considered. Penrose considers that, as the contraction occurs, entropy (disorder) will carry on increasing as time passes so that, when the universe has shrunk back in on itself for the Big Crunch, the entropy will be spectacularly huge, in contrast with when the universe was last so tiny, during the Big Bang. Penrose believes such vast entropy differences can be produced by postulated differences between the space–time structures of the universe at the Big Crunch and the Big Bang.

Hawking's handling of universal entropy makes an interesting contrast. Firstly he notes that he has totally changed his mind on the behaviour of entropy during a possible recollapse of the universe, which serves to highlight the extreme difficulty of the problems involved. Initially he believed that the collapsing universe would be a perfect time reversal of its expansion, so much so that entropy would decrease as the collapse proceeded—a view totally opposite from that of Penrose. Said Hawking, 'People in the contracting phase would live their lives backward: they would die before they were born and get younger as the universe contracted.' Subsequently he decided that this was not correct. Now he believes in another possibility: that there will be 'no strong thermodynamic arrow of time' during a recollapse of the universe. In Hawking's view the universe won't begin to recollapse for 'a very long time'—so long in fact that, because of the hypothesized decay of protons and neutrons into light particles and radiation, 'the universe would be in a state of complete disorder. Disorder wouldn't increase much (during a recollapse) because the universe would be in a state of almost complete disorder already.'

It is obvious that entropy remains an enigma on the cosmological scale. John von Neumann was quite right: so far 'nobody really knows what entropy is anyway'.

CHAPTER 16

TIME ENOUGH FOR OUR UNIVERSE

This journey through the evolution of our understanding of time has taken us from the deserts of Ethiopia to the pyramids of Egypt, through the north China plain, across Salisbury Plain, into Newton's Cambridge, Kelvin's Glasgow, Rutherford's Montreal, Einstein's Berne and Patterson's Chicago. We have seen an extraordinary shedding of ignorance—a genuine widening of horizons. Our accurate knowledge of the age of the earth and our new understanding of our own very recent arrival have given us a perspective against which to view ourselves and our place in the universe, that was totally denied to the overwhelming majority of earlier generations.

And yet, we still don't fully understand time. The origin of the arrow of time is still an unsettled question, for example. In this final chapter, I'd like to look again at the time scale of the universe—its whole past and its distant future. Many cosmologists say that it is impossible to see back past the Big Bang about 15 billion years ago. That in that hellishly hot beginning the laws of physics as we know them did not apply and that it is meaningless to speculate about time before the Big Bang. But we shall now see how curious coincidences among the constants found in the theories of modern physics suggest to us that the universe has no beginning and, presumably, will have no end. The coincidences suggest that the universe has cycled repeatedly through Big Bangs and Big Crunches for an infinite number of past times and will presumably cycle on for ever into a never-ending future. It is reassuring that there is some evidence for this, for how could time begin? How could time end?

130

Figure 16.1: *Blake's Newton—framing the universe.*

The constants of nature

Constants are things that don't change with the passage of time. Yet the study of them has much to say about the very time that they ignore! When you stop to think about it, there aren't many constants in the world. Your mass is not a constant (as most of us know only too well)! Your wealth is not constant (as most of us know only too well). The height of a mountain is not constant when looked at on the geological time scale. The present north star wasn't always perfect north. Even the course of true love never did run smooth.

In their descriptions of the world (figure 16.1), however, physicists have discovered items that do give every indication of being absolutely constant, i.e. they have apparently not changed in value since the Big Bang. They are such things as the masses of the electron and the proton, the charge on an electron, the velocity of light in a vacuum, and the constants that are a measure of the strengths of the electromagnetic, gravitational and nuclear forces. Along with these goes Planck's constant which separates the quantum from the everyday world. These are the real constants of the universe (they determine its frame: the sizes and behaviours of atoms, planets, stars and even the universe itself, ultimately).

One of the first prominent scientists to 'play around' with these constants was the British astronomer Sir Arthur Eddington, whom we met near black holes and who coined the phrase 'the arrow of time'. Eddington was profoundly interested in the most fundamental of problems about the universe and realized (which is still true today) that the ultimate aim of physics is to find all the critical constants needed to describe the universe and, equally importantly, to deduce the exact numerical values of all these constants from fundamental theory. Eddington spent years on the topic and wrote at the popular level with unmatched grace about it but died having failed to convince the rest of the scientific world of the validity of his 'fundamental theory'.

However, Eddington, as in a number of areas, was ahead of his time and in his numerical explorations found that several of the constants of nature have peculiar numerical relationships between themselves. The eminent quantum mechanic Dirac was greatly influenced in this by Eddington and tried to build a cosmology on this basis, although ironically it led him to a theory in which the constant of gravity changes with time! Little came of these theories and for a long period they fell by the wayside, studied only by a few mavericks. In recent years, however, the curious numerical relationships between the constants of nature have become a fairly fashionable field of study and have lead to a contentious concept known as the 'anthropic principle'. To a few physicists the relationships, or coincidences, have suggested the presence of a designer behind the universe. To others they suggest the existence of an infinite number of universes, either all existing in parallel simultaneously and independently or, as we shall prefer, coming successively into solitary existence during the infinity of available time.

What has come to be emphasized more and more in recent times is just how extremely sensitive a universe is to the values of the constants of nature and therefore to their interrelationships. For example, Paul Davies notes that, if the force of gravity were ten times its present value, all stars would burn their fuel ten times more rapidly than they do in our universe. Our own sun would have consumed its hydrogen fuel, expanded grossly in its red giant phase and would have vaporized the earth by now!

In a suggestive study in 1974, Brandon Carter emphasized in a quite different way the extreme sensitivity of the life-style of stars to the constants determining the strengths of gravity and the electromagnetic field. On theoretical grounds he argued that only relatively cool stars would be likely to have planetary systems. A slight increase in the strength of gravity, or a tiny decrease in the electromagnetic field strength, however, would convert all stars into the much hotter blue giants; planetary systems would not exist, and life as we know it could not exist in such a universe.

On the other hand, very slight changes in exactly the opposite directions in the strengths of gravity and electromagnetism would make all stars cooler than now, convection currents would flow more freely in their gases, and they would not explode as supernovae. But all the key elements for life (carbon, oxygen, etc) with the exception of hydrogen (formed in the Big Bang), having been manufactured inside stars, have to escape their factories by means of supernova explosions in order to form living creatures or plants. All the life-forming elements (except hydrogen) would thus be for ever doomed to remain inside stellar prisons, and again no life (as we know it) would appear in the universe.

We can see then the very delicate balancing act that the two force fields are playing for life to emerge in the universe. If either were slightly stronger while the other were slightly weaker, the universe would be lifeless!

Yet another example of the fine-tuning of the constants of nature for life's emergence was given in 1954 by the British astronomer Fred Hoyle, the world authority on the creation of elements in stars. Observing that carbon is the basis for all life, Hoyle highlighted a remarkable feature of this element's formation by nucleosynthesis. For the carbon nucleus to form, three helium nuclei must come together almost simultaneously inside a star. Two fuse together first, followed quickly by a third. For this to happen effectively, the speed of the incoming third particle must be just right. But the speed of particles in a gas is determined by the temperature of the gas, so we are saying that the temperature of the star must be just right for the third helium particle to be welcomed by the other two to produce a carbon nucleus! And, as our existence here attests, the temperature in the star is just right for this carbon production. A similar sort of remarkable coincidence occurs in the production of oxygen. Said Hoyle, 'A common-sense interpretation of the facts suggests that a superintellect has monkeyed with physics...' to get this degree of fine tuning in the universe. And this seems to be the content of the so-called strong anthropic principle—that the constants of nature have been 'chosen' so that life could emerge in the universe.

Such an anthropocentric approach to cosmology, however, is not typical of modern science and is strongly reminiscent of the arguments against Darwin's and Wallace's theory of evolution via natural selection—the argument that, if you see (apparent) design in nature, there must be a designer. The patent obviousness of the occurrence of natural selection, once it had been pointed out, easily demolished that argument from design. There is no need for 'designer genes'. In the present case it's not so immediately obvious how to explain the appearance of clever design without a designer. So far as we know, vast numbers of different species of universes are not spawning new universes which battle against each other in a survival of the fittest Mr Universe competition (although in 1992 Lee Smolin took this concept more seriously in his paper 'Did the universe evolve?' in the journal *Classical and Quantum Gravity*).

However, two ways have been proposed of obtaining designerless design. The first is based on Hugh Everett's many-worlds interpretation of quantum mechanics. In Everett's world, any given momentary situation goes into the future while spawning many new independent parallel universes which themselves are doing the same thing and so on and on. Out of such an avalanche of possible futures obviously any given universe you like can emerge, because all possibilities are being sampled.

A far simpler and much more straightforward way of achieving the same result which has been discussed by John Wheeler is to say that there is no inevitable need for a designer at all, if we allow the expanding universe to brake to a halt under gravity's pull and collapse in on itself again in a Big Crunch, for then the universe could have a new Big Bang. But in this new cycle of expansion the constants of nature could be totally different, so that a completely different universe would expand. It would then stop expanding eventually, and we would get a never-ending sequence of Big Crunches and Big Bangs, with an infinite sequence of different constants and different consequent universes. Given an infinite amount of time, then all sets of constants which allowed bangs and crunches could be experimented with, and ours, of course, would be one of them. Quite probably the overwhelming majority of universes in the sequence would be devoid of life, but that would still leave many that would be teeming with it, and some with life on only a single planet, and so on. On this view we are truly the children of time, infinite time.

Alternatively, when a proper unified theory of the universe has been discovered, one uniting gravity with quantum mechanics, then the values of nature's constants will drop naturally out of the theory, and maybe this will happen in such a way that we'll all say, 'What fools! What other kind of universe could we have expected?'

Index

Printed and bound by CPI Group (UK) Ltd, Croydon, CR0 4YY

22/10/2024

01777625-0020